FIRST PLATOON

ALSO BY ANNIE JACOBSEN

Area 51

Operation Paperclip

The Pentagon's Brain

Phenomena

Surprise, Kill, Vanish

FIRST PLATOON

A Story of Modern War in the Age of Identity Dominance

ANNIE JACOBSEN

DUTTON

Dutton
An imprint of Penguin Random House LLC
penguinrandomhouse.com

LIBRARY OF CONGRESS CATALOGING-IN-PUBLICATION DATA

Names: Jacobsen, Annie, author.
Title: First platoon : a story of modern war in the age of
identity dominance / Annie Jacobsen.
Other titles: Story of modern war in the age of identity dominance
Description: New York : Dutton, [2021] | Includes bibliographical
references and index.
Identifiers: LCCN 2020039784 (print) | LCCN 2020039785 (ebook) |
ISBN 9781524746667 (hardcover) | ISBN 9781524746681 (ebook)
Subjects: LCSH: Afghan War, 2001—Military intelligence—Case studies. |
Biometric identification—United States. | Terrorists—Identification. |
United States. Army. Cavalry Regiment, 73rd. Squadron, 4th. | Military
intelligence—United States. | United States. Army—Stability
operations—History—21st century.
Classification: LCC DS371.412 .J33 2021 (print) | LCC DS371.412 (ebook) |
DDC 958.104/78—dc23
LC record available at https://lccn.loc.gov/2020039784
LC ebook record available at https://lccn.loc.gov/2020039785

Printed in the United States of America
1 3 5 7 9 10 8 6 4 2

BOOK DESIGN BY ELKE SIGAL

For Kevin

Laws are as fair as a spider's webs.
A sparrow will fly through them, but a fly will die.

—JAN KOCHANOWSKI

CONTENTS

PART I

PART II

PART III

PART IV

PART V

FIRST PLATOON

PART I

CHAPTER ONE

THE PANOPTICON

In the summer of 2012, a group of young men, who at the time were American soldiers, found themselves in one of the most dangerous and treacherous places in the world. Thinking they were on one kind of mission, they really were unwittingly part of something much bigger and, perhaps, even nefarious. It was June 4, 2012, and there, on the northern slope of the Arghandab River in Zhari District, Afghanistan, in the village of Payenzai, the soldiers from First Platoon, Charlie Troop, 4/73rd Cavalry Regiment, 4th Brigade Combat Team, 82nd Airborne Division, were getting ready to head into battle, as millions of soldiers had before them, in countless wars across thousands of years. Private 1st Class Samuel Walley remembers the day indelibly, because it was his birthday, and only once in a person's lifetime do you turn twenty years old.

"We captured the number two most wanted Taliban in [the] South," Walley recalled. And though time and priority have diluted the significance of this most-wanted catch, it was important for a number of reasons. "We captured him," said Walley, "by getting his fingerprints."

Private Walley grew up in Georgia on four acres of land, climbing trees and jumping in rivers. His family was military stock. He was raised to be rugged. Turn problems into solutions. In the second grade, he developed a lisp and other kids made fun of him. He willed what was wrong with him away by speaking with a British accent, like James Bond. Now he was twenty years old, six foot, one inch tall. He had green eyes, sandy-brown hair, and weighed 185 pounds. He was fit; he ran a six-minute mile and bench-pressed 240 pounds. Sit-ups were his weakness. He had two atomic bombs tattooed on his left arm with the word "Chaos" written beside them, a nickname from what felt like forever ago but was only last year: high school. His platoon, First Platoon, had been in southern Afghanistan for a little more than three months. Today, Walley carried the biggest weapon on first squad, the M249 Squad Automatic Weapon, a belt-fed killing machine capable of firing 900 rounds per minute. He checked and rechecked his gear. Everyone was ready. It was time to go.

The soldiers lined up in a staggered file formation, originally designed for road marches but adapted for goat paths and grape rows in southern Afghanistan. It was baking hot. Already more than 100 degrees. One of the Afghan National Army soldiers opened Strong Point Payenzai's plywood gate and out the American soldiers went, one after the other, leaving the zigzagged entry control point spaced ten feet apart. The soldiers headed east, then north, into Sarenzai Village, a community of mud-brick buildings and labyrinthine walls located roughly two football fields from the strongpoint. They knew the village well. They patrolled the terrain here two times every day, five or more hours at a time, or until something went wrong.

"Payenzai and everything around it was a hellhole," Walley remembered.

Life here resembled existence after an apocalypse. Decades of war had left Zhari District in a state of collapse. Anarchy and terror had long since vanquished the rule of law. The villagers here had no electricity, no running water, no shops or food stores of any kind. "Mud-brick buildings bombed-out and long abandoned," remembered company commander Captain Patrick Swanson. "No fertile fields, no marketplace, no schools."

"Our minds are dark—we don't know anything, and our children can't even write their names," district governor Niyaz Mohammed Sarhadi told the State Department's Jonathan Addleton in 2012.

At the Pentagon, war strategists called life here "simple and Hobbesian—nasty, brutish and short." State Department estimates put the villagers' life expectancy at forty-four. It was understood by everyone in the American chain of command that the area around Strong Point Payenzai was ruled by a network of Taliban insurgents and that improvised explosive devices, or IEDs, were sewn into the terrain—into the roads and pathways and farmers' fields. Insurgents stockpiled homemade explosives in the abandoned compounds across this area of operations (AO) and recruited new members from among local villagers. Assassinations and murders were commonplace. In 2012, bomb production was on the rise. The Taliban ran IED assembly and distribution across the terrain. The buildings reduced to rubble around here served as bed-down locations for fighters coming in from Pakistan, trafficking weapons and materiel to make more bombs.

After more than a decade of losing this war, in more ways than it could count, the Defense Department believed it had found a technology-based solution to the human problem of insurgency. All across Afghanistan, U.S. military forces were capturing biometric data: electronic fingerprints, iris scans, facial images, and where possible, cell swabs of DNA. The Pentagon believed that through mass biometric collection, it could bring the rule of law to this land where insurgents reigned. The effort was called Identity Dominance. The product was biometrics-enabled intelligence. The battlefield component of this data-driven mission fell to combat infantry soldiers in the field. One member of the platoon, the Company Intelligence Support Team member, the COIST, handled technical collection. The rest of the platoon supported the COIST's efforts with varying degrees of military presence and force.

It was still early in the morning on June 4 as the platoon entered the village of Sarenzai. The temperature had already hit 102 degrees. First Platoon set up a security perimeter. Some soldiers took off their helmets. Others sat down.

"We had the biometrics on maybe one hundred of the military-age villagers in the area," said Captain Swanson, years later in 2019.

"Some of them . . . they had been biometrically enrolled by me on more than one occasion. They were usually working the fields around the area," Private 1st Class James Skelton, the COIST, would later tell a military judge.

Private Walley spotted a man nobody recognized. "He had a white beard, a dark turban, and was maybe forty or fifty years old."

"He looked tired," remembered Specialist Anthony Reynoso.

"I didn't notice him at first," Specialist Dallas Haggard said.

Walley made eye contact with the man. "He was walking in one direction, then after he saw me, he kind of veered off and pretended he was going to the water pump."

Walley gave 1st Lieutenant Dominic Latino a nod. "The lieutenant saw him too."

As platoon leader, Latino was the highest-ranking soldier on the patrol. Off Walley's look, Latino called for an Afghan interpreter, who ordered the man to stop.

Lieutenant Latino began asking the man questions according to protocol. Did he live here in the village? Was he just visiting? If so, for what reason? The man's answers were suspicious. Private Skelton would now use the army's newest biometric collection device, the Secure Electronic Enrollment Kit, the SEEK, to capture the man's biometrics.

Skelton set down his backpack and pulled out the SEEK, a portable piece of military hardware the size of a small shoebox used to electronically capture fingerprints, iris scans, and facial images from civilian villagers and suspected insurgents alike, all across Afghanistan.

Fingerprint matching is a precise science. In order to capture clear, non-smudged friction ridges from a human fingertip, the fingers must be clean. "Most of the farmers' hands were very dirty, and it would sometimes take a package of baby wipes to get a good scan," remembered 2nd Lieutenant Jared Meyer. Scanning each fingertip was an art. The COIST needed to apply slight but consistent pressure for each fingerprint rolled. Thumbs needed to be rolled toward the subject's body, from one end of the nail to the other, but fingers needed to be rolled away from the subject's body, with the knuckle going in, up, and out from the device. The SEEK's collection screen, called a silicone platen, had to be entirely free of smears, dirt, grease, or dryness before each fingerprint scan; the COIST was required to take prints of all ten digits. Uncooperative detainees could be flex-cuffed with their hands behind their backs and fingerprinted that way, but the COIST needed to remember there were different protocols involved when using the SEEK upside down.

Scanning the irises with the SEEK came with its own list of dos and don'ts. Eyelids had to be up, not down. Iris and pupil had to be imaged together with no glare. Direct sunlight rendered the image obsolete. If the person was dead, iris capture needed to happen "within thirty minutes post-mortem," or problems could arise.

Capturing facial images had its own set of technical requirements. A neutral background was necessary, "showing no additional personnel or maps, equipment, vehicles, [or] vessels." Criminal enrollment photos needed to include images of the person from front, right and left profiles, and right and left 45-degree-angled images. "Capture subject's face expressionless and with the mouth closed and eyes open," COIST members were told. And the camera lens needed to be held at the subject's nose height to prevent distortion. There, in this dystopian environment that made Afghanistan one of the poorest, most corrupt, most dangerous and ungovernable nations on Earth, nearly two dozen soldiers of an approximately thirty-man platoon stood in the hot sun waiting and watching while Private Skelton took the man's biometrics. In a war zone, a second can last forever.

After Skelton finished, there was another period of time to wait. The SEEK had the capacity to match new data inputted by the COIST against classified data in the Defense Department's biometrics-enabled watchlist, the BEWL, but there were limitations. The full BEWL was too large to fit on the SEEK, official documentation warned. As a workaround, the Pentagon developed what it called "mission specific BEWLs," meaning watchlists that had been customized for individual terrorists thought to be operating in certain districts or villages.

"Mission specific BEWLs may be loaded onto handheld collection devices," an unclassified monograph reveals, "allowing an immediate alert to be triggered if the sought-after individuals are encountered during identification or verification operations." Which, shocking as it may seem, is precisely what happened next. The SEEK compared the digital fingerprints of the man in the dark turban to the preexisting biometric identities stored on its own drive—and delivered a match-hit.

"The lieutenant did a quadruple take," Walley recalled. "He looked down at his [information]. Checked and rechecked it. The man was the number two most wanted Taliban in [the] south."

The privates flex-cuffed the man. "A woman and a bunch of children came running out of a doorway from one of the compounds," Walley remembered. "One of the kids started beating on me. The woman was shrieking and screaming because we were taking her husband away."

The man was marched back to the strongpoint. From there, he

disappeared into a world guided by the Defense Department's monolithic charter on how to handle detainees. Eventually, he would have been taken north, to a newly renovated American prison next to Bagram Airfield and called the Detention Facility in Parwan. Detractors called this prison Afghanistan's Guantánamo Bay.

Across the 82nd Airborne, the capture of the man in the dark turban was hailed as a success, celebrated at the battalion level, at Combat Outpost Siah Choy, and above it at the brigade level, at Forward Operating Base Pasab. But for the soldiers of First Platoon, catching a middle-aged man at a water pump hardly felt like a battlefield victory in war. And, as it went with so many combat infantry platoons in Afghanistan, this small measure of success was almost immediately overshadowed by a dark turn of events.

"I thought we were in Afghanistan to jump out of airplanes and kill Taliban," Walley told me—looking back through the lens of hindsight, in 2019.

"I thought we were in Afghanistan to kill Taliban and build schools," said Walley's platoon mate, Private 1st Class James Oliver Twist.

"We patrolled hard, knowing we were walking into a minefield every day," remembered Captain Swanson. "From a counterterrorism perspective, we were grasping at straws. We were on a beat, like local cops."

Except they were not police officers, they were soldiers, which made what they were doing less like law enforcement and more like martial law.

Sectioning off an area, registering inhabitants, limiting their freedom of movement, and enforcing compliance with threat of detainment or imprisonment but under the guise of the rule of law is reminiscent of a draconian system of control implemented across Europe in the Middle Ages to fight bubonic plague. This period of time began before the era known generally as the Age of Reason and Enlightenment.

In the 1600s, a series of civilization-threatening plague epidemics gave birth to one of the most heavy-handed surveillance states in the Western world. These were the worst outbreaks of plague since the Black Death ravaged Europe between 1347 and 1351. Urban centers were particularly hard-hit. In London, for example, in a three-month period in 1665, 15 percent of the population died, akin to modern-day New York City suffering 1.3 million deaths over the summer. To combat the epidemic, many villages across Europe were put under quarantine, partitioned into quadrants, and placed under control of a government official. In France, this process was called quadrillage. At a town's gates, an observation post was constructed. Freedom of movement was prohibited; only the government's sentries and administrators were allowed to walk the streets. Michel Foucault, philosopher and historian of ideas, tells us what happened next, in his 1975 lectures and subsequent book, *Discipline and Punish: The Birth of the Prison.*

In order to impose the rule of law, the government ordered the full-scale cataloging of people. A medieval version of biometrics-enabled intelligence. "At the start of the quarantine," Foucault writes, "all citizens present in [every] town had to give their name. The names were entered into a series of registers. The local inspectors held some of these registers, [and] others were kept by the town's central administration." Name, age, and sex of villagers became identifying data for the sick and the healthy alike. Once the infected were identified, these plague-ridden citizens were removed, carted off by government workers. Remaining healthy family members living within each quadrant were required to stay indoors or face punishment, even death.

State-sanctioned rations of food were hoisted into homes by pulleys and baskets. The tracking of people was required; this is what the underlying catalog was for. Every day, twice a day, every member of each family was required to put his or her face in the front window of their home as a designated sentry passed by. "All the information gathered through the twice-daily visits," writes Foucault, was then "collated with the central register held by the deputy mayors in the town's central administration . . . in a kind of pyramid of uninterrupted power."

What this meant was that the location of each individual was to be constantly known to the state, "observed at every point . . . [his] slightest movements are supervised . . . all events are recorded [to include] death, illness, complaints, irregularities." Power was exercised without restraint. "Surveillance had to be exercised uninterruptedly." The state's reasoning for the draconian measures was simple: it had to contain the plague. It had to make sure no newly infected, or sick, were concealed from the state. Hiding an infected person could bring new outbreaks of disease.

With each passing day, each government sentry delivered his data up the chain of command, up the pyramid of power, through the registry. A higher-level administrator turned that biometric data on the people into information about the villages, for the mayor or magistrate to review. Another higher-ranking government official then turned the information into knowledge, in order to determine what action to take next. After a newly infected or dying person was identified, the process of state-sponsored purification began. In some cases, healthy family members were temporarily removed from their home by government law enforcers, the furniture in the house raised up by ropes and suspended over the floor. Windows, doors, keyholes, and cracks were sealed with wax, the home doused with perfume and set on fire. Louis Pasteur's germ theory of disease was still 200 years down the road and officials believed perfumed smoke could get rid of plague. When the decontamination was completed, a curious effort went into play. The government's sentries were body-searched in front of the family, the state's attempt to demonstrate it was only acting in good faith, and that workers could not act with impunity, and had not stolen anything from the home. Martial-law efforts were for society's greater good was the message. An omnipotent government was necessary to rid society of plague. The government was justified in its actions, and was also fair. Of course, perfume did not end the plague, and historians still argue about what, exactly, did.

In the 1970s, Michel Foucault famously drew a parallel between the surveillance state that arose during the time of plague and a prison designed by the English philosopher and social theorist Jeremy Bentham a century after the plague ended. The prison was called the Panopticon, which means "all-seeing." Bentham's architectural plans show a circular prison where all the cells and all the prisoners can be observed by a single guard, watching from a central guard tower. The prisoners, however, cannot see the guard.

The point of the Panopticon was to make everyone feel as if they were constantly being watched. Bentham believed that the fear of being watched could produce the same results as actually being watched would, namely obedience. With the Panopticon, the power of ubiquitous surveillance was that verifiable or not, it still felt absolute.

The idea that the Defense Department's biometric database could become an all-seeing panopticon is not without merit. What follows tells the story of the birth, rise, and weaponization of combat biometrics. How the paratroopers of First Platoon became unwittingly entangled in the Pentagon's efforts is a canary-in-the-coal-mine tale. That one member of the platoon would be convicted of war crimes and sent to a military prison at Fort Leavenworth, Kansas, only to be pardoned by the president of the United States complicates an already tragic chain of events. But there is more to that story, as there is to all of this. This story and its aftermath beg the question, what happens when a federal government is able to tag, track, and follow every one of its citizens, everywhere, all the time?

In 2013, speaking out in opposition to a ruling on *Maryland v. King,* about probable cause and a cheek-swab collection of DNA during a routine traffic stop, U.S. Supreme Court Justice Antonin Scalia warned against the creation of what he called a "genetic panopticon." In Scalia's scathing dissent, he expressed fear for an American future where a person's biometrics could be taken by police, at will, for storage in a federal database for future use.

"I doubt that the proud men who wrote the charter of our liberties would have been so eager to open their mouths for royal inspection," Scalia wrote. The question, then, is: Are biometrics born on the battlefield, without checks and balances, and with questionable legislative oversight, the fruit of a poisonous tree? Are they in the interest of the public they are taken from?

This is a book about identity in the age of identification. About the rule of law in an age of increasing civil disorder and asymmetric warfare around the world. It is a story about how biometric databases are dividing individual people into groups: state versus anti-state, friend versus foe, us versus them. America's Revolutionary War was fought to change "King Is Law" into "Law Is King." What does the future foretell? Is there a new social contract on the horizon, one that crowns biometrics-enabled intelligence a new king?

CHAPTER TWO

THE TWO WILL WESTS

The story begins where part of it ends, at Leavenworth Penitentiary in Kansas. The year was 1903, and an inmate was being booked for admission. For decades, the strange case of the two William Wests was rumored to be a fable. Then, in 1987, Kansas Bureau of Investigation agent Robert D. Olsen, Sr., set the record straight in an article for *Identification News*. It was May 4, 1903, when the storied federal penitentiary in east Kansas received the new prisoner who said his name was Will West, and that he had never been incarcerated at Leavenworth before.

As was protocol for first-timers, the prisoner's Bertillon measurements were taken by the records clerk, a man named M. W. McClaughry, who also happened to be the warden's son. This tracking system—still state-of-the-art in 1903—was based on anthropometry, the science of measuring humans. It had been developed several decades before, by French criminologist Alphonse Bertillon. Bertillon measurements were skeletal-based, meaning body parts were measured by a tailor's tape. Computations like head breadth, length of left foot, and length of the middle finger were painstakingly recorded, then written out on a paper card with a two-image photograph, or mug shot, attached to the record.

Here in the Kansas federal penitentiary at the turn of the twentieth century, the prisoner's measurements card was then placed into the Bertillon System, a card-based database for searching and matching known criminals to crimes by using their unique identifying features on record. The Bertillon System was like a library's card catalog. Entries were alphabetized and searchable by name, but there were also other ways to locate, or cross-reference, information, such as height, or skeletal anomaly like a missing finger or bum leg.

Once McClaughry filled out the card for the new prisoner, he moved to file the card in the Bertillon System. It was at this point that he realized he'd been lied to. In the spring of 1903, there already was a card for William West in Leavenworth's Bertillon System, meaning this man had been an inmate here before. His Bertillon measurements were originally taken two years before, on September 9, 1901, the card said. Sure enough, the mug shot and body measurements were exactly the same.

Here, the story took an unusual turn. The Leavenworth inmate William West was serving a life sentence for murder. What was going on? Had he escaped without notice, gone on to commit another murder, and been caught anew? At once, a guard was sent to check the cell. There, he found William West, inside the cell where he was to be confined for life. Same face, same man. Except it wasn't the same man. There were two William Wests. They were apparently identical twins with the same common-law name.

The following year, the clerk M. W. McClaughry attended the St. Louis World's Fair in Missouri, where he met a sergeant from Scotland Yard named John K. Ferrier. Ferrier introduced McClaughry to a new and improved turn-of-the-century, state-of-the-art technology for identifying criminals. A technology system being used in England and beyond. It was called the fingerprint system. No two fingerprints are alike, John K. Ferrier explained, not even among identical twins.

By year's end, the U.S. attorney general authorized the warden at Leavenworth to adopt the fingerprint system, leading to the Bertillon System's collapse. On September 19, 1905, the ten fingerprints of each of the two William Wests were rolled. To look closely reveals an exclusive pattern of whorls, loops, and arches. Indeed, every set of human fingerprints is unalterably unique.

Human beings have been pushing their fingerprints into the historical record for tens of thousands of years, from the walls of Paleolithic caves 40,000 years ago to clay tablets in the palace in Nineveh, circa 2100 BCE to today. Fingerprint pads are part of every human's individual identity, from before birth. A person's fingerprint ridges first arrange into an infallible means of authenticating individual identity inside the womb, creating a uniqueness that is with us from the moment each of us enters the world until our deaths. No two human fingerprints are alike, no two people the same.

Using fingerprints as a means of verifying identity is traced to the Chinese emperor Qin Shi Huang, circa 250 BCE. By marking royal documents with his own fingerprint, it was as if the emperor was saying, *This is no one else but me.* Archaeology tells us so much about the past, but it also fosters new ideas about the future, which is what happened in 1874, in Japan. A Scottish physician and missionary named Henry Faulds was working on an archaeological dig outside Tokyo when he noticed an ancient fingerprint left behind on a pottery shard. Unlike the emperor's print, designed to demonstrate authenticity, this particular fingerprint looked inadvertent, at least to Dr. Faulds's eye. Imagining how one of these "finger-marks" might be helpful in identifying a criminal, and to use as evidence in a court of law, Dr. Faulds wrote to a colleague, the eminent scientist Charles Darwin, sharing his idea.

Darwin was old and of ill health, and he famously passed the request on to his cousin Francis Galton, an anthropologist, statistician, and eugenicist. Galton had a laboratory at the National History Museum in South Kensington, England, where he was conducting anthropometric studies using the Bertillon System. With this new information, fingerprints became Galton's obsession. "Curved or whorled, having a fictitious resemblance to an eddy between two currents," Galton mused, "the bulb of each finger a maze of minute lineations."

It was the enduring nature of the fingerprint that amazed him most. "From babyhood to boyhood . . . childhood to youth . . . from middle life to incipient old age," to Galton's eye, the bewildering patterns of a man's fingerprint "proved to be almost beyond change." Fascinated with possibilities, Galton began writing articles about fingerprint science and, in 1892, published the first major book on the subject, *Finger Prints.*

In law enforcement circles, Galton's fingerprint science made its way around the world, including to Argentina, where a Croatian émigré named Juan Vucetich served as chief of the La Plata Office of Identification and Statistics. In July 1892, Vucetich would make history. A horrific crime had been committed in the seaport town of Necochea. Two young boys were found brutally murdered in their home, each killed by blunt-force trauma to the head. The devastated mother, twenty-seven-year-old Francisca Rojas, identified the prime suspect to police. The murderer was likely an elderly farmhand named Pedro Ramon Velasquez, she said. Velasquez wanted to marry her, and had even threatened to harm her children if she refused. She was in love with another man, Francisca Rojas told police, she had refused Velasquez's advances, and now her beloved children were dead. Police detained Pedro Velasquez and—so the story goes—interrogated him without mercy, tying him up in a room with the bodies of the children in an effort to elicit a confession. But Pedro Velasquez had a solid alibi and was soon released.

Two weeks later, a fingerprint protégé of Juan Vucetich was dispatched to the scene of the crime. There, the protégé spotted a detail previously unnoticed: a bloody fingerprint on the bedroom doorjamb. It had to have come from the killer, the detective surmised. Francisca Rojas told detectives she never touched the bodies of her dead children. Neighbors told police that when the horrified mother came running to their house after she discovered the bodies, her hands were clean. Juan Vucetich's protégé cut the fingerprint out from the doorjamb and returned to police headquarters with this forensic evidence. The police had a new suspect, and when the mother of the murdered children was fingerprinted, Juan Vucetich got the match he suspected. Confronted with the evidence, Francisca Rojas confessed. Her children stood in the way of her marrying her paramour, she told police, and, yes, she had killed them. Francisca Rojas stood trial and was convicted, the first person convicted on fingerprint evidence used in a court of law. Police stations around the world began replacing Bertillon measurement cards with fingerprint cards.

And a new concept was born: *fingerprints don't lie.*

Using a fingerprint inadvertently left behind at a crime scene opened up a new door in forensic science, a discipline in which professionals analyze physical crime scene evidence for law enforcement purposes. But what about using a fingerprint to single out and identify a criminal hiding in a crowd? What if you could catch a criminal by matching his or her fingerprints against an existing registry or database?

Far away, on the Indian subcontinent, the inspector general of the Bengal Police, a man named Sir Edward Henry, read Francis Galton's book *Finger Prints* and was struck with an idea: fingerprint cards could, and should, be used to create a searchable database of the identities of known criminals. To test his theory, Sir Henry ordered his police force to collect the fingerprints of every criminal in every prison in all of Bengal. He assigned two of his deputies, a statistician named Qazi Azizul Haque and a mathematician named Hem Chandra Bose, to create a mathematical system in which to sort information about fingertips, according to what Francis Galton had codified in *Finger Prints.* This system would become known as the Henry Classification System. It was faster by a power of ten, and infinitely more reliable than the Bertillon System, because no two fingerprints are the same. In no time, the Henry Classification System began replacing the antiquated Bertillon System worldwide, including in the United States.

Fast-forward two decades, when an act of Congress put the Federal Bureau of Investigation in charge of America's fingerprint database. When the database at Leavenworth Penitentiary was handed over to the FBI, in 1924, it contained the fingerprint records of 810,188 criminals, including those of the two William Wests.

FBI Director J. Edgar Hoover was a champion of this new, science-minded, state-of-the-art database. "Through this centralization of records it is now possible for [a law enforcement] officer to have available a positive source of information relative to the past activities of an individual in his custody," Hoover said in 1925. What this meant for the bureau was that a significant new methodology was being systematically pursued for the benefit of law and order. Fingerprints could reliably be used to link crimes from the past to individual persons in the present—criminals whose fingerprints were now and forever in the government's database. FBI clerks assigned to the bureau's Identification Division got to work. More than 7,000 law enforcement agencies around the country began sending fingerprint cards to the FBI's archive in Washington, DC. Over the next twenty-five years, the FBI's fingerprint database would swell: 5 million, 10 million, 11 million cards.

The FBI's public-relations department began promoting fingerprint science every chance it got, including for a *Time* cover story in August 1935. "FBI men reassuringly point out that the bureau's file of 112,500,000 fingerprints [i.e., 11.25 million ten-print cards] were used in some 9,000 cases last year. . . . The FBI got 97.2% convictions." But the sheer number of fingerprint cards the FBI was dealing with had also become untenable. In 1962, a special agent in the Identification Division, Carl Voelker, proposed a radical solution: What if computers could be taught to do some of the grunt work in the time-consuming human job of fingerprint matching?

The FBI turned to its federal partners for help. On the military-science side of the government, over at the Advanced Research Projects Agency (ARPA), great strides were being made in electronic programmable computers. And at the Department of Commerce, inside the National Bureau of Standards, a team of engineers led by scientist Russell Kirsch had recently engineered the world's first device that was able to optically scan images that could be converted into digital data. This new invention was being called a "scanner." What if these two new innovative automated machines could be taught to somehow work together? To read, capture, and save scanned images of criminals' fingerprint cards? What if the FBI's paper database could be transformed into an electronic one?

The challenges were enormous. The scanner had to be taught how to read and record fingerprint images down to Francis Galton's level of minutiae. And these scanned images had to remain consistent over time. These two federal agencies, the FBI and the National Bureau of Standards, joined forces and got to work. After years of development, in 1974, Rockwell International Corporation was awarded a federal contract to build five prototypes of these automatic fingerprint-reader systems. In 1975, the first of these machines was delivered to the FBI. The system was called the Finder. This was the state of affairs at the FBI's Identification Division when a young man named Thomas E. Bush III joined the bureau in 1975.

** ** **

All throughout high school, Tom Bush wondered how he was going to pay for college. Being a kid from a relatively poor family in Mississippi, he got excited when he learned about a scholarship program funded by the Law Enforcement Assistance Administration and created by President Lyndon B. Johnson to fight the so-called War on Crime.

"The deal was," Bush recalled, "every year you worked for law enforcement after graduating from college, they'd cancel twenty-five percent of your debt, which was like three or four thousand dollars at the time." It was the early 1970s, the Vietnam era, and young people were hardly signing up in droves to join law enforcement agencies. Cops still got called pigs, a by-product of the 1968 National Democratic Convention in Chicago, when a group of antiwar demonstrators presented a small pig as their presidential candidate and police intervened.

Shortly after graduating from college, Tom Bush was working a temporary job at a moving company in Gulfport, Mississippi, when he learned the county courthouse was on fire. That's where his high school sweetheart, Cynthia, worked. He rushed over to see if she was okay. As circumstance would have it, standing out in front of the burning building alongside Cynthia was Louis S. Bullard, the sole FBI agent in the small town. The two men began chatting. Tom Bush said he was looking for a job in law enforcement.

"Have you ever thought about the FBI?" Bullard asked Bush.

So began Tom Bush's extraordinary career. He would rise to become the head of the Criminal Justice Information Services Division, or CJIS (pronounced "see-jiss"), the FBI's largest and most powerful division, the bureau's high-technology, identity-information nerve center with an operating budget of more than $1 billion a year. But no one starts out at the top. Tom Bush's career began at the very bottom.

"No one started lower," Bush clarifies.

In September 1975, Tom and Cynthia Bush, newly married, left Mississippi for Washington, DC, in an old Toyota Celica to work for the FBI, as clerks. Tom Bush was assigned to the Identification Division, "flipping and picking" through fingerprint cards. One card after the next, matching a fingerprint on file to a fingerprint pulled from a crime scene took patience, diligence, and, in 1975, the human eye. "It was not a 'lights-out' job." Meaning the system could not operate without a human.

The FBI's Identification Division was massive, with 3,000 clerks working with purpose and precision. "The code of conduct was, 'You do not make a mistake.' If you did, you got three days off," without pay. Another task at hand was scanning the FBI's now 15 million fingerprint cards into the new automated fingerprint-reader system—aka the Finder.

"It was like a big Xerox machine," recalls Bush. "We sat there, scanning cards. All day." The notion that there was simply too much data for human clerks to process at the FBI was about to compound exponentially. In 1976, new provisions to the Freedom of Information Act were passed, including "the right to see records about [one]self," and the FBI was suddenly inundated with citizens' requests for information. In 1977, when Congress learned the bureau had a backlog of 6,000 unassigned FOIA requests on hand, Tom Bush got a fortuitous opportunity. He became part of Operation Onslaught, the FBI's effort to speed up its processing of requests. At night, Tom Bush took accounting classes. Diligent worker that he was, one Friday in 1979, he was given yet another amazing opportunity. "I was told to report to Quantico, on Monday." Moving from FBI clerk to FBI agent was a huge step.

Tom Bush's first job as an FBI agent was with the DC field office, where the majority of its three hundred agents worked white-collar crime. "Just twenty-five square miles, but with so much focus on foreign counterintelligence." But Tom Bush wanted to identify, locate, and capture society's most dangerous criminals, he says, and so, when an opportunity came up to join the FBI's C-4 Fugitive Squad, he raised his hand. For years, he chased bank robbers, fugitives, extortionists, kidnappers, and an array of other dangerous, felony-level criminals. One night, in the winter of 1988, he just so happened to be on "agent duty" on the C-4 Fugitive Squad when his history entwined with the FBI's second most important biometric after fingerprints: facial images, or mug shots.

The FBI had agreed to assist a group of Hollywood producers in their efforts to launch a new, crime-fighting reality TV show called *America's Most Wanted.* The idea was simple: "Watch Television. Catch Fugitives." And the show was popular from the very beginning. As part of agent duty, Tom Bush was working the phones the night that first episode aired.

"The phone kept ringing and ringing," he recalls. Humans are incredibly talented at recognizing faces, and "seventy-five citizens called in and accurately recognized and identified the featured most wanted criminal, from his FBI mug shot shown on TV." The man was a serial murderer and rapist named David James Roberts, hiding in plain sight by using the false name Bob Lord and a fabricated identity. Four days after episode one of *America's Most Wanted* aired, David James Roberts was recaptured and sent back to Indiana's Pendleton Correctional Facility, where he would forever remain.

A human's ability to precisely recognize another human's face has been part of the human condition since humanity began. How, now, to make computers do this? To make automated machines do the grunt work, just as computers had been made to read FBI fingerprint cards? The facial images of criminals were originally hand-drawn, including bank robbers and train robbers on early Wanted posters. The practice of cataloging images of criminals had been around since the popularization of photography. By 1870, the New York Police Department created its first archive of images, called daguerreotypes, which soon became known as mug shots—"mug" being a slang word for face. By 1900, as photograph reproduction became cheaper, the photographs fused with biographic information on criminals, which led to the issuance of the FBI's first Most Wanted Fugitives List, in 1950. These one-sheets were hung in post offices and other buildings, resulting in the occasional tip. But the speed and success of catching criminals by putting hundreds of thousands of human recognizers to work from their living rooms, as *America's Most Wanted* had, was about to be challenged by the idea that maybe machines could be taught to recognize faces too.

The American scientist who first attempted to get a computer to recognize a human face was a nuclear weapons engineer named Woodrow Bledsoe. This was in the late 1950s, when operating a computer required an advanced degree. Bledsoe was interested in automated reasoning. In how to get computers to think. In 1960, with colleagues Helen Chan Wolf and Charles Bisson, he examined a set of law enforcement mug shots. The scientists' seminal idea was to turn those faces into maps. They made coordinates for each human facial feature. Data points for attributes like the center of the pupil, the outside corner of the eye, a point on a widow's peak, the apex of the eyebrow. It was labor-intensive, and also a shot in the dark. This concept had never been tried or tested before. Each scientist plotted out roughly forty facial images per hour, hour after hour, week after week.

Each facial image was assigned a name and a list of corresponding coordinates, all of which was then stored in a database. For the recognition phase, "the set of distances [were] compared with the corresponding distances for each photograph, yielding a distance between the photograph and the database record." The closest-match hits were returned. The machine's pattern-matching capabilities had problems with head rotation and tilt, lighting, facial expression, and age. The scientists kept trying, kept plodding along—trying and retrying—as scientists do. By the 1970s, advances in automated face recognition came from teaching machines to learn nuance: things like hair color, lip thickness, misshapen eyes.

In 1988 semiautomated facial recognition technologies made their way out of the lab and into a police station. The Lakewood Division of the Los Angeles County Sheriff's Department installed a system that could take a mug shot, a sketch artist's drawing of a suspect, or even a still video image of a person committing a crime, and search it against its database of digitized mug-shot photographs. The system sped up the process by narrowing down possible match hits, but a human was still required.

The real breakthrough came in research sponsored by the Defense Department. In 1992, the Defense Advanced Research Projects Agency, or DARPA, the Pentagon's powerful science and technology agency, launched its Face Recognition Technology, or FERET, program, a blue-sky, biometric project. The goal of FERET was to advance face recognition technology research into a single, standardized database for the military, intelligence, and law enforcement communities to share. In 1993, a young, ambitious electronic engineer named P. Jonathon Phillips won an Army Research Laboratory grant that put him in charge of the FERET program's core team. Building on concepts first developed by Bledsoe back in the 1950s, the team finally created an algorithm that could recognize a human face without the aid of a human. To do this, says Phillips, the machines had to be taught to recognize front-facing images, but also how to "perceive" those same faces in left and right profile. The results were promising and the science moved forward, sometimes quickly, sometimes slowly. Still, for another quarter of a century, humans would retain pole position as the so-called super-recognizers.

Around this same time, a third biometric aimed at identifying humans for law enforcement and other purposes emerged, namely the iris scan. Like fingerprints, the iris develops prenatally, in the womb. Although iris color and structure are genetic, the pattern of the iris is unique. And because each human's irises, or irides, are structurally distinct, the iris can be used for recognition and identification purposes. In 1994, the first patent for an automated iris-recognition algorithm was granted. The license and title holder was a British-American computational neuroscientist named Dr. John Daugman. When Daugman took a digital image of his irides in 1993, there was an existent database of one. Twenty-five years later, the number of people whose irides had been scanned for a government database, using the Daugman algorithms for enrollment, exceeded 1.5 billion of the world's 7.5 billion people. India maintains the world's largest biometric database, with fingerprints, photographs, and iris scans on more than 1.2 billion of its citizens.

In the 1990s, the unofficial motto at FBI was "Up or Out," says Tom Bush, meaning you move up the ladder or you leave the FBI. After eight years on the FBI's Washington, DC, C-4 Fugitive Squad, it was time for him to take a field supervisor job. "Jackson, Mississippi, in the 1990s meant violent crimes and civil rights," Bush recalls. The FBI office there was one-tenth the size of the DC office, and he had ten agents under his command. Three years later, it was time to move up or out again. In 1996, he became an assistant special agent in charge of the FBI's Atlanta, Georgia, field office.

As circumstance would have it, he arrived in the middle of a law enforcement national nightmare. A lone-wolf terrorist, later identified as a white supremacist and anti-abortion anarchist named Eric Rudolph, had set off a homemade bomb at Centennial Olympic Park, during the 1996 Summer Olympics in Atlanta. The blast killed spectator Alice Hawthorne and wounded 111 people. A cameraman named Melih Uzunyol died of a heart attack after racing toward the crime scene to help victims. New to the Atlanta bureau, Tom Bush played a minor role in the FBI task force initially assigned to identify, locate, and apprehend the bomber. Then, six months later, in January 1997, another bomb exploded in Atlanta, this time at an abortion clinic in the suburb of Sandy Springs.

The FBI has oversight over abortion-clinic bombings and domestic terrorism. Tom Bush rushed to the scene. He remembers ducking beneath the yellow crime-scene tape, then—another bomb exploded. "I was twenty feet away from Eric Rudolph's second bomb," Bush recalls. Particularly wicked because it was "set up to kill first responders," the bomb was made of twenty sticks of dynamite and flooring nails.

The federal agents at the crime scene were "extremely lucky," Bush says. A lone BMW parked inadvertently between the agents and the bomb absorbed much of the blast. Tom Bush lost his hearing for a while. Seven people were injured, some seriously, including the FBI's Mike Rising and federal agents with the Bureau of Alcohol, Tobacco, and Firearms. "ATF did an amazing job," remembers Bush. "They recovered 70 to 80 percent of the pieces of the bomb. Wire, batteries, explosives, nails, even an egg timer." As he looked at these bomb components retrieved from the crime scene, an idea unfurled in Tom Bush's brain: he wondered if latent prints could actually be pulled from the bomb components retrieved.

A veteran and accomplished ATF agent named Lloyd Erwin said "impossible," Bush recalls. "He explained in great scientific detail how the inferno-like heat and fire from that powerful an explosion would almost certainly have burned off any usable prints." But Tom Bush wasn't so certain. For him, it was a *what-if* moment. For the time being, the two federal agents agreed to disagree.

The following month, the bomber struck again, this time at a nightclub called the Otherside Lounge. Four people were injured. FBI agents landed a fortuitous lead. "We located a second bomb before it detonated," says Bush. But months passed. The FBI had the bomber's fingerprints now, and residual DNA from the unexploded device. But the bomber had never crossed paths with law enforcement before. His biometrics were not locatable in any of the FBI's criminal databases.

Eleven months later, the bomber struck yet again, targeting an abortion clinic in Birmingham, Alabama. Early on the morning of January 29, 1998, security guard Robert "Sande" Sanderson, an off-duty police officer, noticed a flowerpot in an odd position near the entrance to the clinic. As he bent over to have a look, Eric Rudolph detonated the bomb via remote control.

"It blew off Sande's arms and legs," Bush remembers. "All that was left of him was a naked torso." The effort to locate serial bomber Eric Rudolph set off a five-year manhunt, the biggest in FBI history at the time. The search involved all of the high-technology systems in the FBI's arsenal, including helicopters outfitted with advanced cameras and sensor systems. But ultimately, it was an observant citizen who helped the FBI identify Eric Rudolph.

"A college kid recognized his strange behavior near the crime scene," remembered Bush, "and used quarters in a pay phone to call the local police."

In 2000 Tom Bush had been with the FBI for a quarter of a century when it was time again for another "up or out" move. This time the job opportunity was in West Virginia.

He was returning to the Identification Division, the bureau's biggest division. Only now it had gone through a massive information-technology overhaul and been redesignated the Criminal Justice Information Services Division. Tom Bush was its new section chief, Programs Development Section. CJIS was a state-of-the-art data center built on a sloping hillside in Clarksburg. The facility was massive. Its main building was a half-million-square-foot behemoth several stories tall and three football fields long. Its cafeteria seated 600 people; the auditorium held 500. But more important to this story, CJIS was home to the most scientifically advanced tools and services in the nation, designed to help the FBI identify, locate, and catch criminals nationwide and around the world. Inside CJIS's 100,000-square-foot data center, clerks and statisticians compiled and analyzed vast amounts of crime data. CJIS housed two of the FBI's three most powerful criminal databases, the National Crime Information Center (NCIC) and the Integrated Automated Fingerprint Identification System (IAFIS), with the Combined DNA Index System (CODIS), located at Quantico, in Virginia.

In use since 1967, NCIC served the FBI as its electronic clearinghouse of crime data, accessible to every criminal justice agency across America, twenty-four hours a day, 365 days a year. The IAFIS database housed the bureau's automated fingerprint records and was also a searchable system. In 2000, it contained approximately 40 million digital, ten-print fingerprint records, each of which was accessible electronically to law enforcement in all fifty states, as well as some federal agencies. As for CODIS, the FBI's integrated database for digital DNA records, in 2000 it was the most nascent of the automated systems. Begun as a pilot software project in 1990, CODIS was formalized with the DNA Identification Act of 1994. Like the FBI's other databases, CODIS allows state and local law enforcement officers, as well as some federal agencies, to exchange and compare DNA profiles electronically, linking serial violent crimes and criminals to one another through individual criminal profiles.

Tom Bush recalls arriving at CJIS and being utterly amazed. "I'd been in the FBI [for] twenty-five years. I think I know everything. Then I walk into CJIS. A lightbulb goes on," he remembers. "They're running 100,000 fingerprints a day. Very cool. There's NCIC. Just, wow. NCIC is a name-based system," meaning it works to authenticate and certify an identity by matching it to a name. "If I'm going after a criminal named Tom Bush, I want the *right* Tom Bush.

Biometrics allows for that. It lets us make sure we have the *right* bad guy. What I'm seeing at CJIS is something [I've] not seen before: biometrics. What I'm thinking is, *This is huge.* Biometrics is going to make everything different from here on out."

Tom Bush could almost see the future unfolding, right before his own eyes.

CHAPTER THREE

THE HIJACKER'S FINGERPRINTS

Shortly after the 9/11 terrorist attacks, an FBI fingerprint expert and special agent named Paul Shannon was standing in a room inside the New York City morgue when he got a germ of an idea that would change history. It was a grim place, this room. A sign over the doorway, written on yellow legal-pad paper, read: FINGERPRINTABLE BODY PARTS. There had been a terrible crime, a mass-casualty terrorist attack on U.S. soil. There were 2,753 victims who needed to be identified, and yet far fewer body parts. Each person had a unique identity. But who were they? And where to start?

"I would walk with the widows," recalled Shannon. He would point at objects found in the rubble and ask, "Have you ever seen this?" Found objects included wedding rings and neck chains, noncombustible objects able to withstand intense heat or a thousand-foot fall. Something—anything—to help the FBI identify the dead. "So many people had been ground up," Shannon remembered.

In his decades with the FBI, Paul Shannon had never seen

anything like this, and hoped to never see anything like it again. Fingerprintable body parts found in the rubble of the Twin Towers were rare. The New York City Office of Chief Medical Examiner cataloged approximately 20,000 human remains, from which it identified 58 percent of the victims, primarily through DNA testing, not fingerprints. Shannon remembers one success story only: a sixty-year-old man was identified from a forty-two-year-old fingerprint card taken in Las Vegas, Nevada, where he'd been arrested in 1959, age eighteen. Francis Galton was right. From womb to tomb, the human fingerprint stays the same. It was this fingerprint match-hit that gave Paul Shannon the seminal idea.

"Everything about the terrorist attack had been done overseas, with the exception of the execution," Shannon explained in 2019. It was late September 2001. "We had all these people getting ready to head into Afghanistan." The CIA's paramilitary units were already there, and the Defense Department was now sending its own teams of Special Forces operators into battle. "There was going to be a whole bunch of fighters captured on the battlefield. We needed to fingerprint them and we needed to do it to law enforcement standards," Shannon says. It was like a lightbulb going on.

Shannon reached out to CJIS Section Chief Tom Bush and told him his idea.

"My thinking," Shannon explained, "went like this: with fingerprints [in the United States] we can link a burglary in Seattle to a bank robbery in New York. We can find and arrest fugitives. But we can't say who trained at an explosives camp in Afghanistan, and that needs to change. We need to send FBI agents to Afghanistan, and we need to do it now."

When running an idea by Bush, Paul Shannon generally knew when he was on to something from the expression on Bush's face. After hearing Shannon's idea about fingerprinting battlefield captures in Afghanistan, "His left eyebrow went up an inch," Shannon recalled.

Paul Shannon was a field agent. Tom Bush was an administrator who could make good ideas come to fruition by elevating them up the chain of command.

"Write a memo," Bush said.

Shannon sat down and wrote a memo. Bush read the memo and then elevated it to Robert Mueller, director of the FBI. A copy also went to Attorney General John Ashcroft, for legal review.

"CJIS is like an aircraft carrier," Shannon explained. "There are thousands of people running it. Thousands of people operating the machines, making identifications, running fingerprints. At the other end of the spectrum there are people effecting remote identifications in the field. We're like fighter pilots on an aircraft carrier. I was one of them." Shannon requested to be sent to Afghanistan. He wanted to fingerprint the captured prisoners himself.

In his memo, Paul Shannon tried to make the case that the Defense Department was not collecting fingerprints from fighters they detained on the battlefield because "they didn't see the value in it," Shannon wrote. "They were treating everyone they captured like POWs from World War II. As if there'd be an armistice and everyone would go home and tend to their garden in Hamburg or wherever," as had been the case with many top Nazis after the war. "The fighters in Afghanistan were asymmetric warriors. Their idea was, all we have to do is wait. You have children, wives. We have jihad. All we have to do is wait it out. They would tell us this."

Shannon's thinking was that if the FBI had the fingerprints of these battlefield captures on file, in the event of a future terrorist attack, crime-scene evidence could be collected for a potential fingerprint match-hit.

"Be careful what you ask for," said Shannon. "Four days later I was on a plane to Afghanistan."

The plane Shannon flew on was the FBI director's Gulfstream aircraft. In the four days between when Shannon wrote the memo and when the director signed off on it, the CIA's Bin Laden unit read Shannon's memo too. The CIA was in the identity-information business as well, but with a different mandate and a different intended outcome. The CIA is not a law enforcement agency, the FBI is. The CIA cannot arrest people. The FBI can.

On the FBI director's airplane with Shannon was a small team that now included three agents from the Newark field office. Outfitted with printer's ink, 4,000 fingerprint cards, cameras, and DNA swabs, they flew to Pakistan via Italy. "The Gulfstream has these great big windows," Shannon remembers, and he spent much of the trip marveling at the natural wonders of the world down below, including Mount Vesuvius and Mount Everest. "I had to pinch myself and ask, How did I wind up here?"

Before Paul Shannon was an FBI agent, he was a journalist. A "hot spot" reporter for the *Miami Herald* during the city's 1980s crime wave, he covered bank robberies, murders, and other high-risk felony crimes, many of which involved ex-Cuban nationals. Most of these criminals had been former prison inmates released by Fidel Castro and sent to Miami as part of the infamous Mariel boatlift. "Criminals are interesting," he says, "and so are high-risk arrests." Eventually Paul Shannon got recruited by the FBI agents he was using as sources. In 1987, he became an FBI special agent. Now here he was in the FBI director's Gulfstream, about to make FBI history in Pakistan and Afghanistan.

The aircraft landed at a clandestine air base near Peshawar, Pakistan—a former CIA airfield and listening post. This was the location from which the CIA's first U-2 spy-plane missions departed. Renamed Camp Badaber, the airfield was in close proximity to a prison where the Pakistani Army was now housing a group of thirty detainees, men recently captured near the Spīn Ghar Mountains in eastern Afghanistan. They were attempting to illegally cross the border into Pakistan when they were detained. The timing of their capture was significant; it coincided with the Defense Department's bombing of an al-Qaeda cave complex located in nearby Tora Bora. One of the detainees was young, small in stature, unwashed, with a scraggly beard. But he was memorable to Shannon because of the unusual tale he told.

"Most of them had the same story," remembered Shannon. "They were a cook or a driver. This guy claimed he was in Afghanistan to learn the ancient art of falconry. That he only cared about predatory birds." He was nicknamed "the Falconer."

To roll fingerprints, Shannon and the team needed a large, flat surface. The Pakistani Army set up the FBI agents in a room in the prison with an old rickety Ping-Pong table.

"The guards carried [the Falconer] in, strapped down on a kitchen chair. There had been a prisoner uprising on a bus and his legs had been shot up. He couldn't walk. There were no wheelchairs, just a kitchen chair." Shannon rolled the Falconer's ten fingerprints. "I did one hand, first. Then the guards came in and picked him up in the chair, and turned him around so I could do the other hand."

Each detainee was fingerprinted this same way. A mug shot was taken with a 35mm film camera, a DNA specimen collected from the inside of each man's cheek. "Oral swabs like oversized Q-tips were used to collect [these] DNA samples," Shannon recalled. Height, weight, and detainee eye color were cataloged, then handwritten onto fingerprint cards along with alleged date of birth and nationality. Each prisoner was photographed next to a piece of paper or a small erasable whiteboard on which an identifying number was written in longhand. With this, each man's biometric profile was complete.

The team departed Peshawar under the cover of night. "The Pakistani Air Force had set up temporary lights that winked on when we landed, and they winked off again when we flew away," remembered Shannon. "As if we were never there." The next stop was northern Afghanistan, to visit the notorious Sheberghan Prison outside Mazari Sharif. Here, Shannon and the FBI team captured biometric data on each of the military-age males caught fleeing the Battle of Qala-i-Jangi, the first battle of what would be another nineteen years of war. After two and a half weeks of work, that effort was complete. The FBI's goal, to "freeze the identities" of all military-age males caught fleeing the battlefield in Afghanistan, had been achieved.

What Paul Shannon did not yet realize was that his germ of an idea was about to initiate what is known as a revolution in military affairs. It would not only transform future wars, but it would soon reshape the life and liberty of billions of people around the world.

Paul Shannon and the FBI team returned to the United States, hand-carrying the fingerprint cards, photographs, and DNA swabs. The data in these biometric profiles were scanned and readied for upload into any one or more of the FBI's powerful databases.

But there was a hitch. FBI databases are governed by laws enacted by Congress. The biometric profiles hand-carried by Paul Shannon and the FBI team back from Pakistan and Afghanistan were atypical and did not have a legal framework to rely upon. The FBI's existing databases "were comprised exclusively of domestic criminal information," explains Shannon. These new biometric profiles were of people of unknown nationality who had been detained on foreign soil. In order to upload the data on the thirty men caught fleeing the battle space in Afghanistan, and the battlefield captures from Mazari Sharif, the FBI would need permission from the U.S. attorney general to proceed. In March 2002, Attorney General John Ashcroft issued a formal directive giving the FBI the green light to go ahead.

"Fingerprints do not lie," Ashcroft famously declared.

Once the data was uploaded into the systems, the IAFIS computer system began running searches for potential match-hits. The computer kicked back its information, which FBI fingerprint experts scoured and analyzed.

"Delivering a match-hit is not a lights-out process," Tom Bush explains.

In 2002, the most critical roles were for human analysts to undertake. The analyst examining the fingerprint data on the Falconer produced an extraordinary, almost unbelievable match-hit. Paul Shannon was in his kitchen when he learned just how mind-boggling the match-hit really was.

"I received a phone notification," he remembers. "It was CJIS calling. They said, 'Hey, we just got an ID on one of those guys'" in Afghanistan, and referred Shannon to a photograph he'd taken at the prison in Peshawar. Shannon pulled up the image of the man being discussed. "It was the guy that had been strapped to the kitchen chair. Scraggly beard. Bullet holes in his legs. Said he was a falconer." As it turned out, this mysterious, bedraggled pedestrian captured 7,000 miles from America's East Coast was already in the FBI's database. This meant one thing to Paul Shannon. This young man had previously been fingerprinted *inside* the United States by a state or federal law enforcement officer—with probable cause.

"You're not going to believe this," the caller from CJIS said.

The young man's name was Mohammed al-Qahtani. He was a twenty-two-year-old Saudi national. On August 4, 2001, just five weeks before the 9/11 terrorist attacks, Qahtani had been denied entry at the Orlando International Airport in Florida after an astute customs officer suspected he was trying to become an illegal immigrant. When asked if he was meeting any friends or relatives in the area, Qahtani said no. Later, in a secondary interrogation, Qahtani changed his story and said a friend was meeting him at the airport. The customs agent became suspicious of Qahtani's truthfulness and denied him entry into the United States. As part of his voluntary departure, Qahtani's fingerprints were taken and uploaded into the FBI's fingerprint database.

Now this fingerprint match-hit had triggered an alarm at CJIS. Five weeks before 9/11? The Orlando Airport was in the same area where lead hijacker Mohamed Atta lived, trained, and prepared for the terrorist attacks. The federal criminal investigation widened. Airport surveillance video from the day of Qahtani's arrival showed Atta waiting for him outside, at the curb, in a rental car. In 2004, at a public hearing for the *9/11 Commission Report*, Commissioner Richard Ben-Veniste confirmed, "On the basis of that information, as well as significant additional information which we are now not at liberty to discuss in public session, it is extremely possible and perhaps probable that Mohamed al Qahtani was to be the twentieth hijacker."

There is a backstory to consider here. The terrorist attacks of 9/11 were carried out by nineteen hijackers. There were four teams of al-Qaeda operatives who commandeered four commercial airplanes. Each hijacker team had five members—or was meant to have five members—except the team on United Airlines Flight 93, which was missing one man. That this hijacker team had only four operatives is considered a contributing factor as to why that team was overtaken by passengers. Why those passengers were able to force the plane to crash into a field, thereby failing to strike a bigger target—later identified by the commission as possibly "the Capitol or the White House." Prior to this fingerprint match, what had happened to the mysterious twentieth hijacker remained unknown. For the FBI, this one-in-six-billion match-hit delivering incontrovertible identity information on one of the most wanted men in the world was nothing short of extraordinary. For the Department of Defense, the DoD, it was the beginning of a new era.

At the Pentagon, inside the Defense Threat Reduction Agency, DoD's official combat support agency for countering weapons of mass destruction and assessing risk, a paradigm shift was under way. Since the 1950s, threat assessment and so-called risk mitigation focused on the armies and armaments of enemy nations. Watching them from afar to see if they mobilized. Trillions of dollars had been spent on overhead surveillance systems, starting with the U-2 spy plane in the 1950s, spy satellites in the 1960s, and various eyes-in-the-sky platforms through the present day. The terrorist attacks of 9/11 turned all that upside down. The lens went from wide to narrow in a flash. Strategic warning assessment was now, suddenly, personal. Focused not just on armies and armaments but on individual human beings. To be able to identify and arrest a person plotting a crime—as a single fingerprint hypothetically could—might be the difference between a commercial aircraft being hijacked and turned into a weapon of mass destruction, or that aircraft crashing into a Pennsylvania field, or even not crashing at all.

To officials at the Pentagon, the detaining of Mohammed al-Qahtani made clear the idea that individual humans now needed to be evaluated in terms of what would be called personalized risk. In the winter of 2002, no one single person knew precisely how to make this happen. But a new arms race had begun. The methodology would soon have an official name: Identity Dominance. It would spawn countless new ways to wage war.

Change happened fast. The newly created Department of Homeland Security began collecting biometrics from international travelers at all borders and entry points. At the White House, the National Science and Technology Council established its own subcommittee on biometrics. The American National Standards Institute created a Technical Committee on Biometrics to oversee the development of uniformity in standards. The International Civil Aviation Organization drew up a blueprint to integrate biometrics into its automated travel-document-reading machines. Every organization wanted in.

Scores of committees, task forces, and councils of experts sprang up, pondering, then debating, what other biometric "modalities" (that is, other methods) of human measurement to include. Across the federal government, would facial recognition become second to fingerprinting? Or would it be palm-print technology? And what about iris scans? DNA? More meetings, more councils of experts, more plans. More money from Congress. "Senator Byrd got an $800 million, almost $1 billion [sum] for the West Virginia facility," remembered Tom Bush. Defense contractors lined up to win bids.

At the Pentagon, the Biometrics Management Office was tasked with overseeing all military initiatives related to Identity Dominance. The U.S. Army had an early biometrics program to collect fingerprints, iris scans, and facial images that dated back to the Kosovo War and, as per Congress, was to remain the so-called executive agent in charge. Chosen to lead nascent efforts at the Pentagon was John D. Woodward, Jr., a career CIA officer with years of experience as an operations officer in the Clandestine Service and as a technical intelligence officer in the Directorate of Science and Technology.

"FBI was doing fingerprinting in Afghanistan. [The] Department of Defense needed to do its own," Woodward recalled of this time. "There were so many fingerprints to gather. There were significant hurdles to overcome." Woodward understood the power of technology in intelligence, surveillance, and reconnaissance operations from his work at the CIA.

The FBI was there to help. Tom Bush was one of the senior FBI officials leading the efforts, and he recalls this time as a crossroads. Interoperating meant working together. Lack of interoperability, also known as stovepiping, allowed the 9/11 plot to move forward uninterrupted. With all the money and the resources flooding the system, now was the time to act as a unified front.

"After 9/11 our position was, let's find interoperability," says Bush. Acting on behalf of the FBI's largest division, "My goal was to get CIA, MI5, MI6 all connected with interoperability. Connected to 'the server in the sky,' which in 2001 was a new concept. To some [people] it sounded a little like Big Brother. We wanted to be sure we could send a fingerprint card to [the] UK or Australia and they could read it."

No one was blind to the fact that combat biometrics had been born in a time of intense crisis. "[The] Department of Defense had a reactionary build," Bush concedes. The FBI, on the other hand, had eight decades of experience using fingerprints to identify, locate, and catch criminals. To uphold the rule of law.

"When it came to fingerprints there was no one better than us. No one with more experience and technology than the FBI. We are the top dog," says Bush.

"At FBI we know fingerprints better than anyone in the world," Shannon says.

But then disaster befell interoperability efforts. The Pentagon went to war in Iraq.

CHAPTER FOUR

THE BIOMETRIC BELLY BUTTON

It was December 2003 and there was FBI Special Agent Paul Shannon standing with Saddam Hussein, one of the most brutal tyrants of the twentieth century, rolling the dictator's fingerprints onto a ten-print fingerprint card using wet ink. The former president of Iraq had been discovered at 8:30 p.m. the night before, hiding in an underground chamber in the village of Ad-Dawr, Iraq. U.S. forces pulled him out of the hole.

"I'm Saddam Hussein," he told the soldiers who captured him. "I'm the president of Iraq and I'm willing to negotiate."

The man looked like Saddam Hussein. He had a small tattoo on his hand, an initial metric that helped to confirm his identity. But the FBI dispatched Paul Shannon and a small team of FBI agents to capture the dictator's biometrics to confirm that it really was him; he was notorious in his use of body doubles. After rolling his prints, the special agents took a buccal swab from inside Saddam Hussein's cheek, and it was this DNA that gave law enforcement an incontrovertible match-hit. Hussein's two sons, Uday and Qusay, had been killed by a team of U.S. Special Forces and paramilitary operators five months prior. DNA samples had been taken from the dead sons' bodies and uploaded into a classified Defense Department database called Black Helix. The DNA also went into the FBI's Combined DNA Index System, or CODIS. After taking a sample from Saddam Hussein, the FBI's DNA analysts were able to confirm his identity by matching it to the DNA of his two biological sons, from information located on the Y-chromosomes of the male progeny.

While capturing the biometrics of one of the most wanted men in the world, Paul Shannon and the former president of Iraq spoke for several hours. "He fancied himself a poet and an author," Shannon remembers. "He spoke in flowery Arabic, my translator said. He told a lot of stories. In every one he was the hero." Now he was a prisoner of the American federal government.

U.S. forces had invaded Iraq nine months prior. Neither the country of Iraq nor its citizens had anything to do with the 9/11 terrorist attacks. The war was sold to Congress as part of the Bush administration's global war on terrorism, an international campaign to eliminate anti-American insurgencies around the globe. The White House claimed that Iraq's leader, Saddam Hussein, was stockpiling weapons of mass destruction and the Pentagon vowed to destroy these weapons with 130,000 American soldiers deployed in-country. When these U.S. claims proved illusory, a new threat emerged. Al-Qaeda forces regrouped in Iraq, a country in which this terrorist group had no previous foothold. Here began a vicious, bloodthirsty anti-American insurgency, a lethal campaign of roadside bombings and suicide attacks. The Pentagon was blindsided by a new nemesis, the improvised explosive device, the IED. In this mayhem and misery, the Defense Department saw an opportunity for its burgeoning combat biometrics program, starting with fingerprints.

"We wanted to use fingerprints to get bad guys in Iraq," Biometrics Management Office director John Woodward said of this time, "people we could identify as national security threats." The way Woodward explained it through the lens of history in 2019, this sounded like a simple plan. "We wanted to get latent fingerprints off IEDs. Search these fingerprints against the FBI's database and let the computers do all the work." The problem, said Woodward, was that "it's very manually taxing to get latent prints downrange," meaning in the war zone. "But it's important and needs to be done. If you raid a safe house in Mosul, you're going to get fingerprints and you're going to get DNA."

Woodward's idea was a three-step process, he said: "capture prints; send prints to the FBI's database for a search; wait for a match-hit on the prints." In theory, this made sense. Woodward's background, as a CIA clandestine officer, a lawyer, and a former U.S. Army officer commissioned in the Corps of Engineers, gave him a unique understanding of what was needed to accomplish this mission. Adding to his subject matter expertise, Woodward was knowledgeable in the science behind biometrics. He was one of three coauthors of the textbook *Biometrics: Identity Assurance in the Information Age,* published in 2003. But almost everything that Iraq war planners presumed about its military invasion proved false, starting with how Saddam Hussein's army would fight. This presented an intractable problem in the fingerprinting domain.

"The assumption was they'd fight to the death," says Shannon, "which they did not." Instead, Saddam Hussein's fighters laid down their weapons and gave up, en masse. "Nobody was prepared to handle all these fighters as prisoners." Instead of thirty battlefield captures, as there had been in Afghanistan, there were approximately 50,000 Iraqi detainees initially processed by U.S. forces.

"The U.S. Army, which had primary responsibility for combat operations in Iraq, lacked an overall counterinsurgency doctrine," an internal report by the Office of the Secretary of Defense found in 2011. "When one was developed, it did not fully include doctrine related to detainee operations."

The Pentagon scrambled to build prisons in which to house these detainees. To feed and clothe them, and to try to figure out what to do with these individuals next. In the chaos and bedlam, one of the worst scandals in U.S. military history emerged at the Abu Ghraib prison. The prison, built by the British in the 1950s, was used during the reign of Saddam Hussein to house mostly political prisoners, many of whom were abused, tortured, or killed. Shortly before the invasion of Iraq, Hussein opened the gates at Abu Ghraib and let the inmates go. When the U.S. Army took control of the empty prison after the invasion, it was to be used as a detention facility. Lacking oversight and a doctrinal plan, some soldiers with U.S. military police brigades went rogue, committing a series of gross human-rights violations against detainees, including physical abuse, sexual assault, torture, rape, and sodomy. The soldiers took photographs and videos of their crimes, which ultimately led to their arrest. Through all this, the Defense Department worked to capture biometrics on every prisoner detained in Iraq.

Paul Shannon oversaw early efforts at one of these prisons, Camp Bucca, a sprawling detainment facility outside Umm Qasr, near the mouth of the Persian Gulf. Over a six-year period, Camp Bucca funneled more than 100,000 detainees through its concertina-and-chain-link-fenced walls, including Abu Bakr al-Baghdadi, who, after his release, would create the Islamic State of Iraq and the Levant, also known as ISIS, with fellow inmates. At Camp Bucca, Paul Shannon and his FBI colleagues took fingerprints, facial images, iris scans, and DNA of prisoners, one after the next. The data was uploaded into databases, including ones at CJIS in West Virginia, and there it sat. But to what end?

To answer this question, Secretary of Defense Donald Rumsfeld met with a group of civilian science advisers from the Pentagon, a powerful but little-known organization called the Defense Science Board, known among industry insiders as the DSB.

The DSB is a group of civilians who advise military leaders regarding what weapons systems should be built to fight America's wars. Presented in official literature as experts in "science and technology," a perhaps more accurate description is that DSB members are experts in the manufacturing and acquisition of weapons. Many DSB members serve on boards of the same defense-contracting companies that produce the million- and billion-dollar weapons systems that the DSB recommends be built. When Secretary of Defense Rumsfeld reached out to the DSB to discuss the role of biometrics in the global war on terrorism, the question he wanted answered was simple: "Are we winning or losing the Global War on Terror?"

The DSB told Rumsfeld, "The [war] cannot be won without a 'Manhattan Project'–like tagging, tracking, and locating program for national security threats." The Pentagon's biometric profiles on its thousands of Iraqi detainees was a start, but it was passive. This data needed to be transformed into a classified weapons system, DSB proposed, one through which anti-American individuals could be identified for tracking and locating in future operations. The analogy to the Manhattan Project is significant. What began in 1942 as a secret program to build the world's first atomic bomb became the single most expensive engineering program in American history. Instead of ending when World War II was won, the Manhattan Project transformed into its own agency, the Atomic Energy Commission (AEC), the purpose of which was to build more and bigger nuclear bombs. The AEC generated a multitrillion-dollar thermonuclear arms race with the Soviets, with each side creating a vast arsenal of nuclear weapons capable of destroying the world several hundred times over. By the time the Bush administration declared a global war on terrorism, building more nuclear bombs was prohibited by international law. The Defense Science Board saw a new arms race to enter into, and to win, one that focused on mitigating "personalized risk."

When Special Agent Paul Shannon and his team went to Afghanistan, their mission as a law enforcement agency was "to freeze the identities" of fighters caught fleeing the battle space, for law enforcement purposes. What the DSB was proposing to Secretary of Defense Rumsfeld was something different. This Manhattan Project–like program would be far-reaching in scope, a system of systems to track individual people as if they were objects. Analogous to how industry uses tiny electronic tags to track goods in the supply chains. In 2004, this technology system, however, did not yet exist. The Pentagon could build it, the DSB scientists advised.

Rumsfeld authorized the DSB to proceed. In March 2004, the DSB Task Force on Identification Technologies met in closed session at 3601 Wilson Boulevard, Arlington, Virginia, promoting a focus on mitigation of personalized risk. "Our military and intelligence concerns in the Global War on Terrorism have largely shifted away from nation states and their facilities, and toward individuals," official task force literature made clear. Biometric profiles from fighters captured on the battlefield were a starting point, but the DSB proffered that regular citizens needed to be looked at too. Also, according to official documents from the meeting, this meant "assessing personal risk based on threat characteristics such as ideological makeup, social, ethnic, religious and political tendencies."

A second task force group worked on "non-identification technologies," methods with which to tag and track anonymous people, called persons of interest, "over long ranges including [through] electro-optical, radio-frequency, hyper-spectral, and fluid-surface-assembly sensors capable of measuring incremental changes in light, temperature, location." The idea was far reaching, but still farfetched in 2004. Anonymous persons of interest could be watched from overhead, over a duration of time, in order to develop pattern of life activities. The Defense Science Board task force endorsed building an ever-growing fleet of overhead surveillance platforms—drones, aerostats, airships, and other aircraft—from which this new sensor technology could be deployed in order to tag and track people the Defense Department did not yet have identity information on, but would at a future date. Finally, a third task force group worked on the most blue-sky element of the program: "predictive behavior modeling." These automated systems would be designed to gather vast amounts of data on a person of interest's pattern of life, to be analyzed by a computer for predictive-behavior-modeling efforts. In this way, machines would be taught to figure out how to predict what a person might do next.

In 2004, when these ideas were first proposed, this concept was still science fiction. In less than ten years it would become science fact.

To capture biometrics in the middle of a war, the Defense Department needed handheld biometric machines equipped with sensor technology. Thousands of U.S. soldiers and marines could not be walking around Iraq with printer's ink, fingerprint cards, and 35mm cameras. The first piece of equipment fielded was the Biometric Automated Toolset, the BAT, in use since 1999 after the army's Battle Command Battle Laboratory, at Fort Huachuca, Arizona, built a portable system for use in Kosovo. There, U.S. forces regularly hired local nationals to work on its military bases. Some of these local hires got fired for criminal offenses, only to be rehired at other classified military bases in other parts of town simply by using a false name. The BAT was designed as a stopgap solution.

The BAT, if used correctly, could collect and compare fingerprints, facial images, and iris scans. But the tool sets were clunky and had more than a dozen individual components, many with multiple parts. There was a machine for each biometric, plus a tripod, a laptop computer for uploading information, multiple power cords, batteries, and adapters. To understand how to correctly use the BAT, a soldier had to read and comprehend a 103-page instruction booklet. For the war in Iraq, the Pentagon ordered 2,000 Biometric Automated Toolset systems. When a new-and-improved system, called the Handheld Interagency Identity Detection Equipment, or the HIIDE, was developed months later, "lightweight, multimodal . . . and interoperable with BAT," the Pentagon acquired an additional 6,664 devices to be sent downrange, in Iraq.

The public first got a glimpse into the Pentagon's use of biometrics during the battle for Fallujah in December 2004. By then, the Defense Department had expanded its enrollment potential by taking biometrics on regular citizens, starting with Fallujah's quarter of a million residents. "They'll be fingerprinted, given a retina scan and then an ID card, which will allow them to travel around their homes or to nearby aid centers, which are now being built," reporter Richard Engel told audiences watching NBC. The system purported to keep residents of Fallujah safe by targeting people who did not comply. "The Marines will be authorized to use deadly force against those breaking the rules."

All across Iraq, the number and power of IED bombings accelerated. Soon there were near daily mass-casualty IED attacks. Car bombs, truck bombs, human suicide bombers capable of killing dozens of people in a single attack. That an IED costing $25 to make could kill and maim at random, and do millions of dollars in property damage in a single flash, again blindsided the Pentagon. Paul Shannon and a team of fingerprint experts were again forward-deployed to Baghdad, Mosul, and other cities that were being targeted by IED attacks, to help. This time, the goal fell under a counter-IED effort that would become known as "Attack the Network."

"In the beginning," remembers Shannon, "the DoD's attitude was blow and go. Find a device, blow it up, and keep going forward. Our attitude [at the FBI] was, we need to find the bomb makers. If you stop the people who are making the bombs, you are potentially saving tens, hundreds, thousands of lives. Find the bomb maker. Then you can prevent the next device from going off."

Working with various teams of explosives experts in the field, Paul Shannon and the FBI team gathered evidence from IED attacks. "We'd conduct a quick forensic triage of devices and report back to the theater regarding design, appearance, triggering mechanism, and anything else that would help a soldier in the field recognize, and/or neutralize, an explosive apparatus." At the same time, he says, it was clear "the devices could also be exploited in a law enforcement manner for trace evidence—DNA, hair, unique tool marks, explosive analysis, or latent prints." But the reality was, there was no place in the war theater to send any of this fingerprint evidence. "There I was in Iraq with all this evidence. It needs to be processed," Shannon recalls thinking. "So I put it in a box and sent it to the lab at Quantico." One box after the next.

The FBI lab at Quantico was called the Terrorist Explosive Device Analytical Center, or TEDAC. When Paul Shannon started sending bomb components there, the center was just a few months old. Before TEDAC was created in December 2003, there was no single government entity assigned to gather valuable forensics from foreign-made IEDs. No official protocol for exploiting, analyzing, or sharing critical data gleaned from a terrorist's bomb. Back in the 1990s, when Eric Rudolph terrorized Americans with his homemade bombs, the ATF was in charge of the explosives component of the mission. Now at Quantico circa 2003–2004, TEDAC, a multi-agency center, would come together to fit urgent needs overseas in Iraq.

There was a massive construction project already under way to build a new state-of-the-art FBI laboratory at Quantico, Dr. Joseph DiZinno recalls. "TEDAC was to be a subcomponent of this new lab." But the plans had already been drawn, so where to put it? "We built the [TEDAC] lab inside a garage previously used by contractors to receive shipments and park cars. Cleared out a bunch of boxes, wall boards, electrical things" and set up shop. DiZinno, a former dentist, had given up his practice in the late 1970s "in search of adventure and to fulfill a lifelong dream," he says, and joined the FBI. Now, after nearly twenty years with the bureau, he was in charge of the entire forensic analysis branch of the FBI Laboratory—its Explosives Unit, Chem-Bio Sciences Unit, Forensic Science Systems Unit, Trace Evidence Unit, two DNA Units, and more.

DiZinno was legendary among forensic scientists. He'd been part of the original FBI team, called DNA Unit-2, that validated the use of a revolutionary technique to pull mitochondrial DNA from minuscule human remains, like from a single head hair or tooth chip. When Paul Shannon's boxes of IED parts started arriving from Baghdad, DiZinno's new TEDAC team oversaw receipt.

"Our mindset at FBI is that we criminally prosecute," DiZinno says, "so [when] we gather evidence, we do so, so that it will stand up in court." Same with processing evidence in its labs. "Fingerprints, DNA, trace evidence, hair, fibers, tool marks—the collection and preservation of evidence is [the] most important part of the process," he says. "[Anything] not done correctly in the field, and in our labs, means we can't do our job in court." In other words: "garbage in, garbage out."

Precision is time consuming, and the Defense Department saw its needs as more urgent. Despite the fact that the U.S. attorney general had assigned TEDAC the role of federal agency in charge, the Pentagon began making plans for its own forward-deployed forensic labs. In Baghdad, Shannon worked alongside FBI Special Agent Scott Jessee, who was developing a concept called combined explosives exploitation cells, or CEXC (pronounced "sexy"). Initially, CEXC teams sent captured bomb components to TEDAC, though its command post was in Baghdad. But the bomb components kept coming in faster than anyone at the FBI laboratory could deal with. In the first three months alone, more than 800 devices were received at TEDAC.

"In 2005, 2006 . . . we had an eight-to-ten-year backlog of evidence," DiZinno laments. By that time, DiZinno had been promoted to serve as director of the FBI Laboratory at Quantico. Everyone there reported to him.

The Pentagon had by now opened its first expeditionary laboratory, equipped to process all the biometric and forensic evidence it collected in-theater, on its own. Camp Fallujah, Iraq, became operational on January 28, 2006. Seeking additional funding, Secretary of Defense Rumsfeld went before the Senate Armed Services Committee, where he conveyed a false sense of progress. "By March 2006, over 100 latent prints had been received and processed using this new capability," Rumsfeld told Congress. But this number meant nothing without context. In reality, there were tens of thousands of captured bomb components that remained unexamined and unmatched against what would soon be more than 100,000 sets of detainee fingerprints and DNA samples. The Pentagon was moving away from the need to share resources and information with its federal law enforcement partners, and toward transforming its biometrics program into a self-contained weapons system it controlled on its own.

This conflict over autonomy playing out between the FBI and the DoD at the height of the Iraq war had been long brewing, Tom Bush recalls. Recently promoted to the head of the FBI's Criminal Justice Information Services (CJIS) Division, Bush was now responsible for CJIS's $1 billion annual budget. He reported directly to FBI director Robert Mueller, who reported directly to President Bush. For him, the first red flag that something was awry centered on the substandard biometric capturing devices that the Pentagon was sending into the war theater by the thousands.

"DoD's technology was worthless," Tom Bush told colleagues. "BAT, HIIDE, they weren't using a standard." At the Biometrics Management Office, director John Woodward saw many of these same problems. In hindsight, he agrees. "The original BAT was Beta when the world was VHS," Woodward recalls. "There were many problems to overcome," most notably in "standards and interoperability." Just like Tom Bush said.

"The iris scanners, they couldn't do data conversion," Bush adds. "And what's the point of iris scans? People don't leave their irises at crime scenes. Private companies who want to know who is coming in and out of their building, sure. Cool." But with the war raging, "No one on Capitol Hill asked, 'Why the iris scan?'"

Historically within the Defense Department, the separate military services vie for money, power, and control. For their own systems to be designed for their particular service needs. This was the debacle that unfolded next. "First it was the marines," Bush noted. "Then Special Ops guys. Some wanted high-altitude devices. [Others wanted] desert devices, wet-water devices, devices for different climates and different standards. Sure, the guts were the same, but they were not interoperable. Everywhere we looked, there were huge holes."

For Special Operations Forces, the Pentagon purchased 1,648 hand-tailored devices called the Special Operations Identity Dominance Biometric Identification Kits, in two variants. One type, to be used in "less austere environments," cost $10,000 each. The other, for "hold/release info in a tactical environment," cost $2,000 each, for a total of $13.1 million worth of units that could not share information on detainees with its partner forces, or with the FBI. The navy had three biometric collection systems for their use only. One was called the Identity Dominance System, another called the Tactical Biometrics Collection and Matching System, and a third called the Expanded Maritime Interception Operation system for the high seas. These budgets are classified, and these systems are just a few examples of the plethora of essentially useless machines that were forward-deployed.

"It was depressing," remembers Tom Bush. "We'd walk into a facility and see the system with a coffee pot sitting on top."

Tom Bush was not alone in his concerns. "Technologies that allow investigators to identify suspected terrorists have been sped into the field, but these efforts are not being well coordinated, and that can lead to critical information gaps, and so-called stovepipes, the common term for information and communication systems that cannot link to each other," a group of defense advisers at the Biometrics Consortium conference warned. But the Pentagon kept moving forward, pushing biometrics through a series of defense contractor–driven initiatives. Biometrics is "the final layer of proof that no one can spoof," declared Richard E. Norton, executive vice president of the National Biometric Security Project.

When criticized about progress, the responses were generic. "We don't have a single belly button for biometrics in the Defense Department," said Tom Dee, point man for the field in the Director of Defense Research and Engineering Office.

"We are still in the throes of a paradigm shift," insisted Donald Loren, deputy assistant secretary of defense for homeland security integration.

The FBI was tasked by Congress to lead a group on interoperability. "We went to the Defense Science Board to address these problems," recalls Bush. "We felt the problem was giving the authority to the army. That the Defense Department needed its own agency, like a CJIS, to handle [identity] information, and just that." But the request fell on deaf ears. It would take years before a congressional audit revealed just how fallible the DoD's biometric technology programs were at their very onset. Not only did the Defense Department ignore the FBI's biometric standards, but it ignored its own standards—meaning ones the DoD set for itself. "Specifically, a collection device used primarily by the Army does not meet DOD adopted standards," a 2011 Government Accountability Office report found, referring to the HIIDE—the device responsible for "the largest number of submissions collected by a handheld device, according to DOD." It was too little, too late. By the time the report was made public, some of the last U.S. troops were on their way home from Iraq.

There was a second severing of cooperation that would prove even more consequential than unserviceable collection devices. This involved the Defense Department's insistence on having its own biometric database in which to store sensitive and classified data, in what would ultimately become known as the Automated Biometric Identification System (ABIS). The DoD's first iteration, called the Biometrics Knowledgebase System, went online in 2003 and was accessible only to the DoD's Biometrics Senior Coordinating Group and Biometrics Enterprise Working Groups. These individuals "have been fine-tuning the site's technical elements since then to incorporate user feedback and publish the initial stages of device testing and evaluation methods," said the director of the Biometrics Fusion Center, army major Stephen Ferrell, calling the database a "Website" protected by Secure Sockets Layer. And that soon it would be "accessible to all government users with .gov or .mil addresses." Interest was so nominal, there were just 100 registered users. "DoD anticipates 10,000 user registration requests a year," predicted the Biometrics Fusion Center's spokesman.

"But why did DoD need its own system?" Tom Bush asks rhetorically. "This was a question people all across government were asking. Why a separate system? Plainly stated, it was mission." After 9/11, there were three key organizations dealing with this new business of biometrics, the FBI, the Defense Department, and the Department of Homeland Security. "Each with a very different mission," Bush clarifies. "DHS deals with good people. 99.9 percent of their travelers are good people. Their hit rate is super low. At FBI, we deal with bad people. One hundred percent of the people in our criminal database are there because they've been involved in the criminal justice system. DoD had a different need." The DoD fights wars. "DoD does what it does." And what the DoD does with its database—whom it collects and keeps information on, and why—is mostly classified.

The original goalposts set up after the 9/11 terrorist attacks had inalterably shifted. John Woodward's three-part plan, developed at the Defense Department's Biometrics Management Office in 2003, involved using the FBI's powerful IAFIS database for match-hits, in partnership with the DoD. "We found al-Qahtani this way," Paul Shannon recollected, in an interview in 2019. But with the creation of its proprietary ABIS database, the Defense Department changed course—entering into a domain it knew its FBI partners were prohibited from going into. ABIS is a searchable system that stores biometric data the FBI cannot legally store. When the FBI captures biometric data from a person, a series of standard operating procedures are in play, including probable cause. When the DoD wants biometric information on a person, it acquires it: sometimes by force, other times by manipulation, but almost always without that individual comprehending that their highly personal information will likely be forever at the Defense Department's fingertips.

"With ABIS . . . DoD started collecting things on people we [at FBI] are not authorized to collect," Tom Bush explained in 2019. "The privacy people [in the United States] would go insane."

On September 23, 2004, the DoD's Biometrics Management Office awarded defense contractor Lockheed Martin a five-year contract to design, build, and maintain ABIS. It was the beginning of a $3.5 billion budget for combat biometrics over the next eight years.

The ABIS database began expanding at an astonishing rate. In the summer of 2007, the Biometrics Task Force reported having 1.5

million records. By 2008, that number had nearly doubled, to include biometric profiles on more than 2.7 million Iraqis. But as the DoD had done with its biometric collection devices, its ABIS program surged forward with a colossal budget and little oversight. Not for another seven years would an auditor from the office of the Director, Operational Test & Evaluation (DOT&E), who reports directly to the secretary of defense, assess ABIS's operational capabilities as "not operationally suitable . . . cumbersome . . . inadequate [with] deficiencies exist[ing] in the areas of training, usability, and Help Desk Operations." Perhaps most troubling of all, the auditor determined, "ABIS v1.2 is not survivable against unsophisticated cyber threats." But instead of correcting these egregious problems, the DoD moved forward with a new audacious plan, one that was astonishing in scope and control. The most powerful military force in the world would now begin collecting biometrics on almost the entire population of an unwitting foreign country, Afghanistan.

In 2010, the U.S. Army's newly created Task Force Biometrics, Afghanistan, announced it had begun gathering biometrics on 80 percent of Afghanistan's estimated 25 million citizens. The two-year goal set by the Defense Department was 21 million sets of fingerprints, iris scans, and in some cases DNA, starting with all military-age males. By the fall of 2011, the Defense Department had biometric data on more than 2 million people in Afghanistan. Data it had uploaded into its proprietary ABIS database.

Also in the fall of 2011, a *New York Times* reporter named Rod Nordland went to Kabul to report on this largely unknown program. In service of the story, Nordland, "an American of Norwegian rather than Afghan extraction," allowed the army to capture his biometrics with its BAT system. After his fingerprints and iris scans were "entered into the B.A.T.'s armored laptop," Nordland wrote, "an unexpected 'hit' popped up on the screen, along with the photograph of a heavily bearded Afghan."

The Pentagon's ABIS database had misidentified Rod Nordland as Haji Daro Shar Mohammed, an Afghan man on Terrorist Watch List 4: "Deny Access, Do Not Hire, Subject Poses a Threat."

One look at Nordland's American passport made clear he was not Haji Daro Shar Mohammed in disguise. But this false positive was indicative of a deadly array of problems. These problems included, but were not limited to, gross misidentification by a classified automated system sold to Congress as infallible. So much could go terribly wrong. And would.

PART II

CHAPTER FIVE

GEOGRAPHY IS DESTINY

It was February 2012, and the young paratroopers of First Platoon were heading into the war theater in Afghanistan. Their first stop was the U.S. Transit Center at Manas, Kyrgyzstan, a landlocked country of 6 million people in Central Asia. The country had been part of the Soviet Republic, America's former nemesis, since 1936 but regained independence in 1991 after the USSR was dissolved. In 2001, the American government made a deal with the Kyrgyz Republic, creating this strategic air base to function as a staging ground for men and materiel headed into war. Over the next eighteen years, more than 775,000 American service members would deploy to Afghanistan at least once.

Geography is destiny, it is said. That where a person is born invariably influences what happens to them in their life. For those who sign up for the nation's ultimate service, the terrain into which you deploy can also guide your fate. The area in southern Afghanistan where First Platoon was heading was the Taliban's home territory. They had dominion here, sovereignty and control. If all of Afghanistan was on fire, and metaphorically it was, then Zhari, Panjwai, and Maiwand Districts were at the center of the inferno. In the staid military-speak of the Defense Department, these three districts made up the "focus of insurgent activity" in 2012.

First Platoon flew into Manas on a charter flight, a 747 non-military aircraft with regular seats, its overhead bins crammed with M4 carbines, M249 Squad Automatic Weapons, M320 grenade launchers, and more. "Bolts out, zip-tie in the chamber so they can't fire," says Staff Sergeant Daniel Williams, one of the platoon's non-commissioned officers. Each soldier had a rucksack, two duffel bags, and an assault pack for overnight missions. Sergeant Williams carried a red poker chip in his pocket, for good luck.

The aircraft touched down and the soldiers deplaned. The sky was gray and it was snowing in Kyrgyzstan. Growing up in Georgia, Private Walley rarely saw snow. He almost wasn't on this deployment, which is curious to think about, considering what would happen to him here. During pre-deployment training at Fort Bragg, Walley got singled out as a troublemaker, broke his left ring finger screwing around, and almost didn't deploy. "The army needed bodies," he says, and he went. A short while before leaving for war, Walley remembers standing in the bathroom at Bragg, looking at himself in the mirror and feeling self-conscious. Not entirely clear of the man he wanted to be.

"The army forges you a new identity," Walley says. "It takes certain goals away and gives you new ones. For some people this is good, for [others] it is bad. Very few of us stay the same." Looking back at the time in Afghanistan, during a philosophical discussion on war, in 2019, Private James Oliver Twist agreed. "You go in one way and come out something else."

Moving across the tarmac at Manas, Twist walked alongside Walley. The two of them would be assigned to the same weapons squad. Walley was heavy weapons and drove armored vehicles.

Twist carried the Thor, an electronic, counter-IED jamming system invented for the war in Iraq. Twist was shorter than Walley by a few inches, with brown hair that acquired a reddish tint when filled with Kandahar dust. He wore thick-lensed eyeglasses and wished he didn't, especially when night-vision goggles (NVGs) were involved. "That's how I was born," he said. "Some things you can't change." Private Twist and Private Walley were both nineteen when they deployed, their birthdays less than a month apart in June.

Twist wanted to be a soldier from as early as he could remember, and now he was one, which made him feel proud. One of his first sets of drawings, in elementary school, depicted the Battle of Gettysburg in crayon. Twist grew up in Michigan near a pumpkin patch and a horse farm. As a kid he practiced soldiering in the backyard, moving G.I. Joes in formations around the sandbox and shooting cap guns from high up in trees. He was named after an uncle Oliver, not the character in the Charles Dickens novel, but the family still had fun with the high-minded jokes. They were like that, the Twists—celebrated Twistmas at Christmas and that kind of thing. Some boys join the army in high school to escape terrible home situations, including divorce, violence, and poverty. Not James Oliver Twist. "I was running toward war for some other reason," he recalled.

Private Twist was in the fourth grade when the hijacked airplanes hit the buildings on 9/11, and that image seared into his brain, he said. In family photographs from that time frame he wears camouflage outfits and can be seen waving the American flag in his small hand. "It was a direct line from 9/11 to the 82nd Airborne," said Twist. To becoming a U.S. Army paratrooper with the legendary airborne infantry division.

Twist's father, John, served in Vietnam. Both grandfathers in World War II. A British great-great-grandfather served in World War I but never saw action because he was stationed in Halifax, Canada. In October 2009, when James Twist was sixteen, he accompanied his dad to his Vietnam War platoon reunion. It was held at the Hilton Hotel in Grand Rapids, and in the spirit of military pride James Twist wore his father's army uniform from 1968.

"It fit him perfectly," John Twist recalled. "In the elevator going up, a stranger turned to him and said, 'Thank you for your service,'" apparently mistaking him for an active-duty service member.

"You're welcome," James Twist said in response, as if sending himself a message back from the future, through time. Both father and son felt great. Shared a smile.

During his senior year in high school, Twist's mother, Caroline Scott Robinson, got cancer and died, and his thinking took a left turn. For the first time he could remember, he began to seriously question the meaning of life. "I needed a purpose. Something to focus on, or things could get dark," he said. In notebooks, Twist doodled his thoughts, and his mother's initials, "CSR," on nearly every page. He wrote and rewrote one of the last things she said to him before she died: "Don't let yourself get weak. You are as strong as you stay." In the spring of 2011, Twist walked into a recruiter's office and joined the U.S. Army. He was eighteen years old.

"I was born for service," he told them. "I was created for it."

During pre-deployment training at Fort Bragg, Twist kept a journal and wrote in it almost every day. Considering what would happen to him, the first entry is ominous and prescient:

> August 3—Last night I dreamed of a deadly car accident that involved many important people in my life. I dreamed that Christian Heumpfurer died. I made a phone call to his mom.
>
> "Hey James," crying.
>
> "Is this really happening?" I stated.
>
> "Yes."
>
> End of conversation.
>
> I ran to a party that same day and talked to people. They seemed content with his death. Like they were al-

ready over it. This tells me that I take lives and good friends really seriously. I love friends and family [and] if any were to die I'd have some more bad days. I feel like the Army was a good decision but also in mind is a lot of dark thoughts about Afghanistan or Iraq.

—I could die

—I could come back with PTSD

—I could be massively injured and live with a disability

—Friends could die

I have so many more thoughts about the war. Maybe it will start winding down soon. I have many more bad dreams . . . This place seems like a purgatory of some sort . . . I will try my best. I am as strong as I stay. R.I.P. CSR.

Snow was falling hard as the soldiers made their way across the tarmac, heading toward the Transit Center facilities in Kyrgyzstan.

"Walking with all your gear was a balancing act," recalled Specialist Todd Fitzgerald, one of the platoon's marksmen. When Fitzgerald was a kid, people made fun of his anatomy, of his tall and lanky body. "They called me the skinny stick," he remembered. After high school, he was motivated to get his airborne wings and prove them wrong. Show that you could "be tall and skinny and get straight A's" and make an excellent soldier too.

Between the weapons and the packs each of them carried, and the ice on the ground, the tarmac was slippery.

"One guy ate it and fell over," Fitzgerald remembered. "Someone yelled 'sniper!' and everybody laughed. You see, army humor is dark humor. There was a point in time where we could all still make light of things." That was then.

Private 1st Class Zachary Thomas, still eighteen and endowed with an ingenious map-reading skill called orienteering, was standing near the soldier who fell over. Thomas laughed, exhilarated by the thrill of going to war.

"When we were in Manas it seemed fun, still," Thomas recalled. Being from Crosby, Texas, he, too, was unfamiliar with snow.

After dropping their gear in the barracks, many of them went outside to participate in a snowball fight against other units heading to war.

"Army versus Navy," Walley joked.

Private Thomas and some other soldiers made a snowman with an army cap for a top hat. They took photographs to send to family members back home.

During President Barack Obama's first years in office, he surged 33,000 troops into Afghanistan, raising the number to more than 100,000 come 2010. The following summer, the president addressed the nation to say troops were starting to be withdrawn, first 10,000 troops, then 23,000, then another 10,000 after that. With most soldiers heading home, First Platoon's paratroopers were like firefighters rushing into the World Trade Center to help.

"The army trains you to go where you're needed, and we were needed so we went, full-on," Sergeant Williams said. But for what, exactly? Back in the fall of 2008, an out-of-the-ordinary plan of action set in motion by the previous administration of President George Bush would begin to take effect. After seven years of bloodshed and relatively few gains, the Defense Department had decided to change tactics in Afghanistan.

"We cannot kill or capture our way to victory," Secretary of Defense Robert Gates told an audience of national-security strategists at the National Defense University at Fort McNair in Washington, DC, in the fall of 2008. The new tactic was going to be about imposing the rule of law across Afghanistan. Four years later, in 2012, First Platoon was part of this effort.

"Without ROL [rule of law] the country cannot progress no matter what contributions are made by outsiders," the State Department told the Defense Department in a now declassified report called *Rule-of-Law Programs in Afghanistan*. This conclusion was supported in a series of monographs and reports that all said essentially the same thing. "The absence of a modern, functional government sustains the Taliban and Al-Qaeda," the inspector general warned, stating the obvious. The rule of law requires a criminal justice system and Afghanistan did not have one. "The infrastructure of the system of law enforcement, courts, and corrections had been eradicated," the State Department found.

What existed in Afghanistan was a dispute-resolution system that was not acceptable to the U.S. government, Department of Defense officials told Congress in a follow-on report. Sociologists, anthropologists, criminologists, and national-security experts sent to Afghanistan reported back on rule-of-law issues and explained that traditional Afghan dispute resolutions were "inconsistent with international human rights principals [*sic*]." One example highlighted was "the practice of *baad*: A traditional practice of settling disputes, primarily in Pashtun tribes, whereby a young girl is traded to settle a dispute for her older relatives."

Congress supported Defense Department efforts to fix the problem. "Establishing the rule of law (ROL) in Afghanistan has become a priority in U.S. strategy for Afghanistan and an issue of interest to Congress," the House and Senate agreed. Because the country had "[no] infrastructure system of law enforcement, courts, and corrections—prisons and jails," the Defense Department would spearhead the creation of a new, Western-style criminal justice system there. Starting in 2009, a series of multibillion-dollar programs were initiated, all based on the rule of law. The glue holding these endeavors together was biometrics.

International Security Assistance Force (ISAF) commander General Stanley McChrystal requested White House approval to establish Joint Task Force 435 (later renamed Combined Joint Interagency Task Force 435), with its goal "to centralize detention operations, interrogation, and rule of law functions in Afghanistan." McChrystal's deputy commander was Brigadier General Mark Martins, a Rhodes Scholar with a degree from Harvard Law School, and a former judge advocate with the U.S. Army. General Martins would put together an organization called Rule of Law Field Force-Afghanistan.

Whereas General McChrystal looked at the battlefield from the top down, General Martins's organization handled operations from the ground up, "providing field support to rule of law efforts." In the system of law enforcement, courts, and corrections, the combat infantry soldiers would act in the law enforcement role. Biometrics efforts surged: 7,000 Biometric Automated Toolset (BAT) systems were fielded to Afghanistan, and 12,000 soldiers were trained in how to capture fingerprints, iris scans, facial images, and DNA in a war zone. Biometrics were sold to Congress as a panacea. A cure for all ills.

In the fall of 2010, the army's Task Force Biometrics held a press conference in Kabul to explain how, and why, its biometric efforts would work. There had not been a national census in Afghanistan since 1978, U.S. Army colonel Craig Osborne said, "So identification of its lawful citizens is problematic." Osborne outlined the Pentagon's top three goals: "To help our Afghan partners understand who its citizens are; to help Afghanistan control its borders; and to allow GIRoA [the Government of the Islamic Republic of Afghanistan] to have 'identity dominance.'" Stated succinctly, these goals were:

— Identification of citizenry
— Border control
— State authority

Most Afghan citizens received almost nothing of value from the state. A majority of the rural population, which was estimated to be 85 percent of the people, received no social services from the government of Afghanistan, American investigators found. Most of these rural areas had no electricity, no running water, no sewage systems, healthcare, police, or security. The biometric cataloging of people purported to change all that, Osborne announced. "Knowing who its citizens are will enable the Afghan government, in the future, to provide services to its people in a precise and effective manner," he said. Who could argue with that idealistic premise? Basic services are a core tenet of the social contract, the theoretical agreement between the people and the state.

It was in the late seventeenth and early eighteenth centuries, as the plague epidemics came to an end, that modern ideas of social-contract theory became refined. One of the basic tenets of the social contract in civil society is that an individual must give up some of his or her rights in order for the social contract to work. Society's members cooperate for the greater good, and in doing so, they receive benefits. The philosopher John Locke famously argued that the primary right that individuals give up in order to live in a civil society is the right to punish other people for violating rights. There can be only one criminal justice system in a civil society, namely the one run by the state. This is why law enforcement, courts, and corrections are the backbone of civil society. You cannot just go around killing or extorting people, or trading a young female family member to an older relative for sex, as a means of settling disputes. Not in a society based on the rule of law.

To help create Afghanistan's criminal-justice system, there were Afghan partners involved in the Pentagon's biometric efforts, Colonel Osborne announced. The Afghanistan Ministry of Interior, Biometrics Center, located in Kabul, was run by Afghan lieutenant colonel Mohammad Anwar Moniri. The Afghan-led element of the program, also funded by the Americans, was called the Afghan 1000 and it would begin soon. It would be staffed by "Afghan enrollers and [Afghan] system administrators [who] will collect biometric data at key border locations and in concert with the Afghan Population Registration Department offices across Afghanistan," according to Osborne. The short-term goal was to capture ten fingerprints, two iris scans, and one facial image on 80 percent of Afghanistan's 25 million people. The longer-term goal was to have "GIRoA assume full control and ownership of biometrics" with no date set, but to occur as soon as possible.

In 2010, most people were entirely unfamiliar with the concept of combat biometrics, and during the army's press conference, Colonel Osborne took a few minutes to explain how the science worked. "Biometrics allows Afghan and ISAF forces to precisely target terrorists and insurgents that are operating in Afghanistan. The technology allows us to gather fingerprints and DNA, for example, from things such as documents or unexploded IEDs and match identities to negative events. For example, if an unexploded IED is found along a road, we are able to develop latent fingerprints from that device and then compare them to known identities in our database to determine who had touched the IED."

At the end of his speech, Colonel Osborne turned the microphone over to Colonel Moniri to say a few words.

"The goal of the [Afghan] Ministry of Interior Biometrics Center is a secure Afghanistan," Moniri said. "[To] separate law-abiding Afghans from insurgents and criminals."

Getting criminals off the streets would benefit everyone, was the idea. In America, requiring an entire population to turn over highly personal biometric data would, almost certainly, be contested as a Fourth Amendment violation—an unacceptable transgression of a citizen's right to privacy. But this was Afghanistan. It would be years before Americans learned these same military surveillance methodologies would eventually come home, to tag and track citizens in the United States.

In Kyrgyzstan, the facilities at the Manas Transit Center had all the amenities a combat infantry soldier could ask for on his first deployment. "The DIFAX [dining facility] was open 24/7. You could get a candy bar from the vending machines at two A.M.," said Fitzgerald, whom everyone called Fitz. That evening, on one of the overhead TV screens near the entrance to the gym, there was breaking news on CNN. Riots in Kabul.

"Tires and things on fire," remembered Walley.

"Shit hitting the fan," remembered Twist.

"Some shortsighted individuals burned [some] Korans," Fitz recalled.

Making matters even more complex for rule-of-law efforts in Afghanistan, the Korans that had been burned came from the largest American-run prison in the country, the Detention Facility in Parwan. In the paradigm of law enforcement, courts, and corrections—that is, prisons and jails—this American-built prison was supposed to be the correctional system's jewel in the crown.

"I remember thinking, Really? Someone actually did that?" Fitz recalled. "I'm from Texas. Can you imagine a bunch of people from a foreign country showing up and burning Bibles from your local Baptist Church? How stupid can someone actually be?"

The Koran burnings would have deadly consequences across Afghanistan, setting off a disastrous chain reaction, like dominoes falling. The events would create a political firestorm in Washington, DC, and directly affect First Platoon beyond anything they could have imagined at the time.

CHAPTER SIX

KABUL IS BURNING

The holy books at the heart of the ruinous Koran-burning controversy were from the prison library. The Detention Facility in Parwan had recently gone through a multimillion-dollar U.S. government–led renovation. It was the largest prison in Afghanistan and was located roughly one mile east of Bagram Airfield, the largest U.S. military base in-country.

The old prison, called Bagram Collection Point, was built during the Bush administration and had been plagued by an ugly criminal past. In 2002, two Afghan prisoners were beaten to death there by U.S. Army soldiers. One of these men, a cabdriver known as Dilawar of Yakubi, was later determined to be innocent of any crime. He'd been falsely fingered as a terrorist by a corrupt Afghan warlord working for the Americans. The warlord turned the innocent man in for a $1,000 Rewards for Justice–style payment, criminal investigators later found.

The new prison in Parwan sought a better image. Constructed

with Geneva Conventions requirements in mind, it featured three 950-bed housing units. Unlike Afghan jails, the one in Parwan had an electrical system, sewage system, and drainage system. "The facility offers extensive medical care and classes in literacy and trades such as agriculture, bread-making and tailoring," according to an army press release. "Families are allowed to visit detainees and the facility includes a playground for the detainees' children." In keeping with rule-of-law efforts, the Pentagon wanted to send a message. Afghanistan's criminal suspects were to be taken care of while they waited for a fair and impartial trial. This facility housed violent jihadists as well as many civilian villagers caught up in the military's dragnet. Guards kept track of detainees from catwalks. There was high fencing, klieg lighting, and watchtowers. "This [facility] is part of the overarching goals for the war, to capture the insurgents and Taliban and concentrate them in one area," project manager Harry Pham explained.

Next door was the Justice Center in Parwan, the courts component of the criminal justice system being set up by the Defense Department. By 2012, there were 58 U.S. justice advisers training 110 Afghans how to prosecute Taliban bomb makers using forensic science. Cases were built almost entirely on biometrics, on fingerprints and DNA pulled from bomb components, as Colonel Osborne had laid out at the press conference the year before. The 110 Afghan personnel included officials from the Ministry of Justice, the Attorney General's Office, the Ministry of the Interior, the Supreme Court, and the Ministry of Women's Affairs. The detention facility and the justice center had been built to work hand in glove. That was the idea.

The Koran-burning incident began at the detention facility on the evening of February 20, 2012, when prison guards with the 42nd Military Police Brigade, an army police unit headquartered at Joint Base Lewis–McChord, in Washington State, learned from one of their Afghan interpreters that prisoners were passing information to one another by writing notes in the margins of prison library books. The books at issue included 474 copies of the Koran and 1,123 other books deemed "extremist literature" by the interpreter. "[He] characterized the books as Nazi-like texts, encouraging extreme interpretations of Islam," an internal army investigation later found.

The army's battalion commander at the prison ordered his soldiers to get rid of the library books in question. Removed from the shelves, they were boxed up, loaded into armored vehicles, and driven to Bagram Airfield down the road. There, a little after 3 A.M., two uniformed soldiers began throwing the Korans into a burn pit for incineration. To destroy or debase the Muslim holy book is generally perceived as an aggressive and offensive act. There were several Afghan workers present at the burn pit, and when they realized copies of the Koran were set to be burned, they became upset and objected. When the American soldiers refused to stop, the workers got physical.

"We attacked them with our yellow helmets, and tried to stop them," a laborer named Zabiullah told the *New York Times*. But it was too late. Several Korans had already started to burn. Zabiullah and his colleagues threw water on the books and surreptitiously pulled them from the fire as evidence of a crime. Out of view of the American soldiers, the Afghan workers videotaped images of the charred pages and uploaded them to the Internet for all to see.

By morning, more than 1,000 people had gathered outside Bagram Airfield to protest and express rage. By nightfall, the number swelled to 10,000 and the protest transformed into a riot. Men and boys hurled rocks and incendiary devices at guard towers. The crowd burned effigies and set tire barricades on fire. When tower guards at Bagram Airfield fired rubber bullets into the crowd, the violence got worse.

"When the Americans insult us to this degree, we will join the insurgents!" an enraged man told BBC News.

These were not Taliban insurgents, journalists learned, but

mobs of Afghan citizens. After more than ten years of war, Afghan citizens were fed up with American power and authority over their lives. President Hamid Karzai condemned the incident. The Ulema Council, a government-funded body of religious clerics who advised the secular government on matters of Islamic jurisprudence, met in Kabul. "There are some crimes that cannot be forgiven, but that need to be punished," declared council member Maulavi Khaliq Dad. The Afghan clerics had their own way of settling disputes, and due process was not one of them. The Taliban called the Koran burning an insult to "one billion Muslims around the world." U.S. Secretary of Defense Leon Panetta said what happened with the library books was "inappropriate and deeply unfortunate."

In Kabul and Washington, DC, there was rancor at the highest levels. Presidents Karzai and Obama were now officially at each other's throats. Karzai threatened to pull out of the status-of-forces agreements. Obama ordered his top general, ISAF commander General John Allen, to go on national television and formally apologize. "We are taking steps to ensure this does not ever happen again," General Allen said in a televised expression of regret, apologizing to numerous groups and individuals. "I assure you, I promise you, this was not intentional in any way, and I offer my sincere apologies for any offense this may have caused. My apologies to the president of Afghanistan. My apologies to the government of the Islamic Republic of Afghanistan. And most importantly, my apologies to the noble people of Afghanistan." General Allen looked wooden and his words sounded rehearsed. No U.S. official wanted to admit just how far the two countries were apart in terms of what defined the rule of law, let alone how to enforce it in Afghanistan.

From the dining hall in Manas, the soldiers of First Platoon marveled over the stupidity involved. How preventable it all seemed.

"In retrospect, it is easy to pass judgment on problems that don't affect you," Twist said in 2019. "When they affect you" it is a different story, he said. "You can [groan] at how insensitive the world is."

The riots killed thirty people and injured hundreds more. The lethal fallout did not end there. It continued on just days later, in a wave of retaliatory killings called green-on-blue attacks. This is when an Afghan soldier or policeman turns on his American counterpart and kills them, usually pulling out his American-issued weapon and shooting what was supposed to be a trusted colleague at close range, often in the head. This terrifying reality sowed distrust and cynicism across an already fragile alliance.

From Manas Air Base, the paratroopers flew to Kandahar Airfield in southern Afghanistan, where they received grim news. Two of their fellow soldiers from the 82nd Airborne had been killed in one of these revenge attacks. Staff Sergeant Jordan Bear and Private 1st Class Peyton Jones were standing in a guard tower at a combat outpost in Zhari District when they were ambushed and killed by an Afghan National Army soldier and an Afghan literacy teacher.

Staff Sergeant Michael Herrmann, another one of First Platoon's noncommissioned officers, or NCOs, took it especially hard. He and Jordan Bear were platoon mates on an earlier deployment, in Arghandab District, Afghanistan. "A lot of things that happen on a deployment become one big blur," Herrmann said. "It's hard to separate out events. That day I remember for sure. Within a day of learning Bear got shot in the head, I learned my grandfather died. I was pissed. I was in no mood for war."

But soldiers do not get to take days off. They learn to compartmentalize the horror of war, and to ignore the difficulties and tragedies happening back home. They press on.

At Kandahar Airfield, counter-IED training had a different focus from training at Fort Bragg, which emphasized post-blast triage. "Things like how to tie a tourniquet on a pig with a blown-up limb," Private Twist recalled, and then remarked on how after the war he rarely discussed these things. "People really focus on the cruelty-to-pigs part of it, and leave out or ignore [the fact] that this can happen to your friends."

Counterinsurgency specialists set up an obstacle course. First Platoon's paratroopers practiced ways to avoid IEDs, learning the Pentagon's latest in-theater counter-IED tactics and techniques. "They set up little mazes to practice how to walk through a minefield without stepping on a mine," remembered Sergeant Williams.

The Taliban's IEDs were crudely constructed bombs built to deceive. Engineered from innocuous-looking items like plastic bottles or oil containers—anything disguised to look like trash. The IED was a quintessential asymmetric weapon. Like green-on-blue killings, it bred paranoia and dread. That regular Afghan citizens driving cars, trucks, motorcycles, and bicycles to get to where they needed to go could also be insurgents driving car bombs, truck bombs, motorcycle bombs, and bicycle bombs into a target to produce a mass-casualty attack was debilitating. "You had to shut your mind off and do what you had to do," said Twist.

After Kandahar, the platoon helicoptered west, to Forward Operating Base Pan Kalay, located in Maiwand District. Like much of the terrain across southern Afghanistan, the area around Pan Kalay was wracked with violence and anarchy. Weeks before First Platoon arrived here, the Taliban had assassinated the deputy provincial governor, two local policemen, and two of the governor's sons—a poke in the eye of rule of law. The army launched Operation Creature Pan Kalay in response, sending soldiers knocking down doors and clearing compounds in search of bomb factories and homemade explosives. The operation netted 400 pounds of IED-making material, only to soon be replaced by hundreds more pounds of the same thing.

At night company commander Cpt. Patrick Swanson sometimes read *Winning Insurgent War*, a counterinsurgency treatise by a former U.S. Army lieutenant colonel named Geoffrey Demarest. The book had been assigned by 82nd Airborne brigade commander Colonel Brian J. Mennes for all the officers to read.

"All of the battalions got boxes of them," Captain Swanson recalled. "Required reading for E-6's and up."

In an insurgency, such as the one in Afghanistan, "Justice and Liberty are held hostage by sinister actors," Demarest wrote, "control over events and power bypass questions of legitimacy (rule of law), defaulting to whomever acts with orchestrated impunity." It was fancily written and needed to be unpacked, Swanson told me. Impunity, exemption from punishment, was a loaded word.

At twenty-seven years old Captain Swanson had already served two tours of duty in Iraq, and had been awarded the Bronze Star for meritorious service in combat. Now he was in charge of approximately 100 soldiers: First Platoon, Second Platoon, and the Headquarters Company, which together were called Charlie, or Chainsaw, Troop. As for many other soldiers, Swanson's call to action was 9/11. He was at boarding school in Washington, DC, when the planes were hijacked. One of his friends, a foreign exchange student from Japan, had an apartment with a rooftop view and when Swanson learned the Pentagon had been hit, he ran over to his friend's place to try to get a visual on what was going on at the U.S. seat of military power. The two high school students stood there, watching black smoke rise from a gash in the building.

"That did it for me," Swanson recalled.

He applied to West Point and got in. After graduation, his first deployment was to Iraq, where he spent months fighting the insurgency war. The desire to win was as motivating a force to him now, in 2012, as it was when he signed up for service after 9/11. He believed in the army and its mission. He followed orders from his superiors, as captains do, and he used his own skills, as a company commander, to lead the men who followed his. Such is the nature of the chain of command. The pyramid of power.

On the first page of *Winning Insurgent War*, he read, "All warfare is influenced by topography but to best comprehend some conflicts, we need to look closely at the interrelationships of people with their surroundings." In the theory of counterinsurgency operations, anonymity and terrain are always entwined, and with deadly results. "Anonymity is that obscure quality of going unnoticed or unidentified, which helps an insurgent or outlaw move without being caught," wrote Demarest. It was anonymity that allowed the Taliban to move freely around Afghanistan's most volatile southern districts, here in Maiwand, Panjwai, and Zhari, and to control this terrain. Biometrics, wrote Demarest, were the answer to unmasking the terrorists, seizing power, and taking control.

"If at all possible, give everybody an ID card," Demarest proffered, but "biometrics can't be allowed to mean just careful physical identification of perpetrators, or of someone from whom we want to deny access." Biometrics had to be captured from every person living in every village, to form a central registry, like in a plague town. "Biometrics has to be applied to everyone for Big Brother to work as well as it might," Demarest wrote, the reference to Orwell apparently in jest. To impose the rule of law across Afghanistan, to win this insurgent war, the state needed first to know who was who and who owned what. "A peaceful liberal social contract is inextricably dependent on formalized public records of ownership," he wrote. To his eye, Western-style rule of law was set up to protect what people owned.

But in southern Afghanistan, the IED was law, or king. It was the centerpiece of Taliban-style law enforcement. The insurgents used IEDs to punish and terrorize soldiers, policemen, and villagers alike. Farmers were coerced into allowing Taliban to bury IEDs on their land. The Defense Department's overhead surveillance systems regularly spotted Taliban bomb emplacers making arrangements with farmers, teaching them how to disarm a daisy chain of IEDs before farming their own land, provided they arm the bombs again after finishing a day's work.

IEDs were ubiquitous, they were everywhere. One of the nightmare elements in 2012 was that Taliban insurgents had mastered a way to defeat many of the counter-IED devices, like the Thor that Private Twist carried on his back. In Iraq, IEDs were mostly triggered by cell phones. By the time the Pentagon's Joint Improvised Explosive Device Defeat Organization came up with its array of electronic jammers, the insurgents in Afghanistan had moved on. Here, now, the new nemesis was the pressure-plate improvised explosive device, the PPIED, a homemade bomb triggered by the weight of a person's limb. PPIEDs were generally constructed without metal. Insurgents packed agricultural components, like ammonium nitrate fertilizer, into plastic vessels and buried them in the sand. The army's top countermeasure against the PPIED was the human eye. Across Afghanistan, soldiers trained as explosive ordnance disposal, or EOD, specialists learned to spot these death traps with a refined combination of vision and experience at staying alive.

At Pan Kalay, First Platoon was joined by EOD specialist Staff Sergeant Israel P. Nuanes, the most seasoned soldier in the platoon. At thirty-eight years old, Nuanes was old enough to be many of the privates' father. He'd enlisted in the army in 1992, the year many of them were born, and began his army career as a truck mechanic deployed to Iraq. After seeing the carnage that IEDs could unleash on the bodies and minds of his fellow soldiers, Nuanes volunteered to become an EOD specialist. On a second deployment to Iraq, Israel Nuanes earned the Bronze Star for heroic service. Now he was a staff sergeant. "He was so good with the younger soldiers," recalled Sergeant Herrmann. "He taught them how to spot IEDs and they looked up to him."

"He could find an IED with a stick," like a medicine man, remembered Walley.

"And he had an enviable mustache," recalled Twist.

"We always joked about what we couldn't wait to eat when we got back home," Herrmann remembered. "Mine was a burger. His was tacos with the hottest hot sauce [in New Mexico] where he was from."

EOD specialists are the army's preeminent tactical and technical explosives experts, made famous by the Hollywood film *The Hurt Locker*. The position in which they serve the nation is, arguably, one of the army's most dangerous jobs. As per official U.S. Army literature, there is no "safe" procedure for disposing of IEDs, "merely a procedure that is considered the least dangerous."

"I always thought EODs wore bomb suits," remembered Walley. "In Afghanistan they did not. Nuanes would walk around with his sleeves up, teaching us things that would save our lives. He showed us the whole country was a minefield."

In Pan Kalay, the patrols began immediately.

"Nuanes, the EOD, always out front in the lead," recalled Twist.

"Nuanes told us, 'Stop trusting the equipment,'" Walley recalled. "Equipment fails. He taught us: to reach supreme accuracy always be on the lookout with your eyes. If rocks are stacked together, that's an indicator of an IED. Piece of tarp poking out of the dirt? Another indicator. Taliban leave clues so local villagers can spot IEDs, especially kids. If a kid gets blown up, the Taliban loses hearts and minds too."

First Platoon had been in Pan Kalay for a few days when the unthinkable happened. One of the most horrific acts ever committed by a lone U.S. soldier unfolded just ten miles down the road, at a small army outpost, called a village security platform, in Belambai Village. The events would become known around the world as the Kandahar Massacre.

CHAPTER SEVEN

MURDER, MAYHEM, AND CONSEQUENCE MANAGEMENT

At 1:05 A.M. on March 11, 2012, in the bitter cold and the deep dark of night, an Afghan army soldier on guard duty at Village Security Platform (VSP) Belambai, in Panjwai District, saw a shadowy human figure exit the outpost on foot. There was no fortified gate controlling access into the outpost, just a wooden pole wrapped in barbed wire that moved up and down, allowing armored vehicles to come in and go out during the day. It would be easy for a soldier to duck under the pole and leave the facility, but no American ever left VSP Belambai alone, let alone at night, and the Afghan guard told criminal investigators the reason he did not report the unusual activity right away was because he thought his eyes were playing tricks on him. This remote U.S. military garrison thirty miles west of Kandahar sat in the middle of an ungovernable area, where there was no rule of law. The Green Berets stationed at VSP Belambai were tasked to set up a local police force.

"[We] have been unable to stand up an Afghan Local Police force," an army officer conceded in an official report.

Six days prior, on March 5, an armored vehicle on its way back to VSP Belambai with five American soldiers inside drove over an IED just a few hundred yards from the garrison's front gates. The blast flipped the truck, gravely injuring several of the soldiers. EOD technician John Asbury was part of the quick reaction force unit that raced out to assist the men trapped inside. Approaching the overturned vehicle, Asbury either stepped on an IED or it was detonated remotely. The ensuing blast amputated Asbury's left leg. One of the soldiers who stayed behind, pulling base security, was a staff sergeant named Robert Bales. He later described how watching what happened to Asbury, through the scope on his sniper rifle, left him feeling helpless and enraged. This region was filled with insurgents who wanted to kill you, Bales told investigators. Nowhere was safe. It made sense when, a few days later, the Afghan guard on duty thought his eyes were playing tricks on him when a figure appeared to slip outside into the darkness. To leave VSP Belambai in the middle of the night made no rational sense.

Fifteen minutes after the guard saw the lone figure exit the garrison, he heard gunfire coming from the north, near the village of Alikozai. People in this village were so poor, had so little property or prosperity, they lived alongside their cows. The gunfire stopped. Minutes later, the Afghan guard watched as a man approached the garrison and walked back inside through the front gate. The guard could now see the man was an American soldier, he said, because of the night-vision goggles he wore and the assault rifle he carried. At 2:30 A.M., a second Afghan guard on night duty observed a soldier leave through the front gate. He reported what he saw to the highest-ranking soldier on post, a Special Forces captain named Daniel Fields.

It was below freezing outside, a thin layer of frost covering the ground. The Afghan villages around the base had no electricity. Everything was dark and silent until a single motorcycle began heading out of Alikozai, its headlight swinging across the mud-colored structures, the abject poverty, everywhere the sense of war. Driving the motorcycle was a villager named Habibullah Naim. Behind him, in a borrowed truck, his brother, Faizullah Naim, was transporting five critically wounded gunshot victims lying prone in the back. Habibullah Naim was leading the way to Forward Operating Base (FOB) Zangabad, he later told investigators, because like many of the villagers in the area, he had "taken wounded persons [there] in the past." Faizullah pulled up to the gates, stopped the vehicle, and cried out for help.

"An American shot them!" he told the Afghan guard, pointing to the bloody victims in the back of the truck.

FOB Zangabad had a warning system for mass casualty attacks, and an air horn was blown to alert medical personnel to come to the aid station at once. Inside the trauma bay, two battlefield surgeons, asleep on cots, leapt up and began receiving the five casualties. "The first patient that arrived was a young girl," one of the surgeons recalled, "estimated age around six or seven years with a gunshot wound to the head. The likelihood that she would be able to make a full recovery was low, so we didn't want to spend our time treating her when we knew we could save the others."

Down the hall, FOB Zangabad's tactical operations center, called the TOC, spun into action. A second girl, a teenager, had a gunshot wound to the chest, but the oldest male victim, her father, refused treatment for her because their religion required a female to be present, he said through a translator. As army personnel in the TOC scrambled to bring in a female medic from FOB Masum Ghar down the road, the medics tended to a small boy who appeared to have also been shot in the head. "The bullet nicked the pinna, the outer part of the ear, causing a laceration." The oldest man, Mohammed Naim, now understood to be Habibullah and Faizullah's father, had been shot in the neck and jaw. He was "remarkably awake and alert and had a GCS [Glasgow Coma Scale] of fifteen and was communicating with the interpreter," the doctor told investigators. The fifth victim, a teenage boy, shot in the left thigh and the right hip, was also alert and awake. As the army medics treated the five victims, an interpreter worked with Faizullah to try to understand what had happened. By 4:10 A.M., information had moved up the chain of command and the TOC at Kandahar Airfield's Camp Brown was making arrangements for medical evacuation.

It was 3:15 A.M. at VSP Belambai when Cpt. Daniel Fields ordered an accountability check of all soldiers on base. With roll call under way, some of the soldiers reported hearing gunfire to the south, near the village of Naja Bien. At 3:20 A.M., Fields learned an American soldier was missing from his cot; it was Staff Sergeant Robert Bales. Captain Fields's first thought: "[I] worried he'd been kidnapped."

Every Army fort across southern Afghanistan was supposed to have an outward-facing surveillance camera called a Cerberus, named after the three-headed beast that guards Hades in Greek myth. At Belambai, the Cerberus surveillance system was broken and awaiting parts. At 3:35 A.M., Captain Fields called the battle captain at FOB Zangabad to request help locating Staff Sergeant Bales using their state-of-the-art overhead surveillance system called a Persistent Ground Surveillance System, or PGSS (pronounced "pee-jiss"). Developed for the war in Afghanistan, the PGSS was a giant aerostat, a lighter-than-air balloon. Flying at an altitude of approximately 2,000 feet, the balloon carried an array of high-definition imaging systems, including electro-optical and infrared cameras. The PGSS was one element of what the Defense Science Board proposed in 2004, part of its Manhattan Project–like tagging, tracking, and locating program. In March 2012, there were roughly sixty aerostats in Afghanistan, approximately a dozen of which were flying over southern Afghanistan, designed to keep constant surveillance over the dangerous stretch of road along Highway One from Kandahar to Helmand, and the villages all along the Arghandab River to the south.

Captain Fields asked the PGSS operator at FOB Zangabad to focus on an area just south of VSP Belambai, where soldiers heard gunfire. To be searching for an American was an unusual and high-priority task, a DUSTWUN, the acronym for "duty status, whereabouts unknown." "The person [we are looking for] may have been sleepwalking," Fields told the aerostat operator at 4:15 A.M. Working the keyboard and screen at FOB Zangabad, the technician scanned the area while soldiers at VSP Belambai fired 60mm mortar illumination rounds to assist with the search. Chief Warrant Officer 2 Lance Allard and a quick reaction force readied themselves to leave the outpost and look for Staff Sergeant Bales.

At FOB Zangabad a female medic finally arrived from FOB Masum Ghar and began attending to the teenage girl who had been shot in the chest. "[We began] placing chest seals over the penetrating chest trauma as well as doing chest needle decompression," she told investigators. Next, the medical team turned back to attend to the littlest girl, shot in the head and clinging to life. "She had a 2- to 3-centimeter hole in the top of her head where brain matter was exposed," the surgeon told investigators. The girl's body was tiny. There was no pediatric laryngoscope equipment in the aid station. The surgeons struggled to intubate her with equipment designed for adults. The endotracheal tube was too large, and failed. The girl's oxygen levels dropped dangerously low. One doctor pushed medications while the other worked to restore her breathing. Finally, success.

The Afghan interpreters had by now determined that all five victims were civilian casualties, or CIVCAS. The acronym CIVCAS presented a dreaded situation for any commanding officer, not just for humanitarian reasons but for legal ones. For doctors, though, the CIVCAS designation aided in their efforts. CIVCAS meant priority use of a nine-line MEDEVAC, in this case to transport the victims to Kandahar Airfield by helicopter instead of slow-moving armored vehicles. The airfield in Kandahar had a world-class trauma center, a NATO Role 3 Multinational Medical Unit. But the winter weather was bad, and deteriorating quickly, and all Black Hawk helicopters at the airfield were grounded. "I recommended the army use the Pave Hawk helicopters like the pararescuers do [because] Pave Hawks can fly in low visibility," one of the surgeons recalled. The highest-ranking officer at Camp Brown's tactical operations center authorized the request. The Pave Hawk was on its way to Zangabad. All five victims would survive.

"But the [younger] girl has never spoken since," BBC Pashto reporter Mamoon Durrani said in an interview in 2019.

Back at VSP Belambai, Captain Fields worked with the PGSS operator, who, together with a second colleague, was now scanning the area south of the outpost. Using thermal sensor technology meant the PGSS operators were searching for a heat signature on the ground. At 4:36 A.M., the operators spotted what appeared to be a human form crouched down in a farmer's field.

"I saw the heat spot," one of the operators told criminal investigators. "I said, 'Go back. [There] is a body that is moving to[ward] Belambai.'"

The second operator moved the camera in close on the shape that was radiating heat: a man lying in a prone position, "as if trying to hide." The crouching figure "appeared to be carrying a rifle and had NVGs [night-vision goggles]. He was wearing some type of shawl." The man got up and began moving. The PGSS operators used the camera to follow him as he walked north toward the military outpost, avoiding what was understood to be an IED-sown road, always staying in the field.

"About halfway to VSP Belambai, [the soldier] began running

until he arrived at the VSP," the PGSS operator told criminal investigators.

CW2 Lance Allard and his quick reaction force team were about to exit VSP Belambai when they saw Staff Sergeant Bales emerge from the darkness. Allard remembered how it went down. "While at the front gate we noticed an infrared strobe coming from the west. We began to call out to SSG Bales and he began to respond back." Allard ordered Bales to stop, and to lower his weapon.

Bales stood still in the darkness. He was close enough that his fellow soldiers could see clearly that he was covered in blood. "He had this cape thing [on], multi-cam pants, and I think he had a T-shirt on," Allard told investigators. "He had his helmet on his head and that is how we picked him up and saw him." Bales was searched. The cape was a piece of fabric he had stolen from one of the victims' homes. "I discovered that he had multiple weapon systems, many covered in blood," Fields told investigators. Staff Sergeant Bales was disarmed, then moved into the operations center, where he was placed under guard.

The Defense Department's Village Security Operation program was a key component of its counterinsurgency efforts. Sold as a way to build "atmospherics" and "rapport" with local villagers. Across Afghanistan, these programs had been failing since they began. At VSP Belambai, soldiers were fired upon the first day they arrived. They continued to receive direct fire "every time they went out," according to declassified reports. "These guys are going on patrols to engage with the community, and [every day] on the way back [they] are fighting to get to their home base."

In 2012, there were ninety Village Security Operation sites across Afghanistan, each fort built in a remote and highly contested area. Run "in conjunction with counter-terrorist and counter-guerrilla operations," the mission was "to extend the reach [of the] Government of the Islamic Republic of Afghanistan and create an environment inhospitable to insurgent influence." A rule-of-law support effort. But in three months of steady patrolling, there had been no measure of success. When Bales went AWOL from the outpost and murdered what would turn out to be sixteen civilians, there was no local law enforcement in the area to investigate the crime. The Afghan police assigned to it would come from Kandahar City, metaphorically a world away.

The Green Berets running operations out of VSP Belambai were assigned to Special Forces Operational Detachment Alpha 7216. They were supported by an infantry unit, in which Staff Sergeant Bales served. "The infantry [being] the 'game changer' that permits a larger and more robust presence," according to army paperwork. At VSP Belambai, Staff Sergeant Bales was the senior member of the infantry uplift squad. His assignments included base defense, guard duties, and camp improvement and expansion. He also went on combat patrols in support of the Special Forces soldiers stationed there and helped train Afghan National Army soldiers, known as the ANA. Age thirty-eight, Bales had joined the army after 9/11. In 2012, he had served in the army for nine years, three and a half of which were spent in combat. This was his fourth deployment in the war theater.

Under guard in the VSP's operations center, Bales looked dazed as he sat on a bench soaked in his victims' blood. "We asked [Staff Sergeant] Bales what had happened from start to finish," Captain Fields said. "Bales stated that he did not wish to tell us anything about what happened. He basically plead[ed] the fifth." He asked if he could take a shower and change his clothes; both requests were granted. After washing himself clean and dressing in new clothes, Bales asked for his computer, an HP laptop. After a soldier handed it to him, Bales threw it on the floor and crushed the screen with his boot. Investigators later determined Bales had been watching videos of war dead.

The sun rose over Kandahar Airfield. It was now 6:23 A.M., three hours after the five gunshot-wound victims had been dropped off at

FOB Zangabad's gates. The U.S. Army's Criminal Investigation Command, known as CID, officially opened its investigation. In one of its official logs, a lieutenant colonel whose name remains redacted "reported an American soldier killed and/or injured several Afghan local nationals near Village Stability Platform (VSP) Belambai."

The CID is the primary criminal investigative arm of the U.S. Army, the organization within the Department of Defense that investigates felony crimes and other violations of military law. It is often called the army's FBI. The CID supervisor assigned to the case was Matthew Hoffman, a civilian police detective from Phoenix, Arizona. Hoffman and his colleagues were now working as Army Military Police reservists in southern Afghanistan. "Unfortunately, we are the army soldiers who police the army soldiers," said Hoffman in 2020.

Seventeen CID investigators were assigned to work alongside Special Agent Hoffman. This was a criminal case of unprecedented scope, a cold-blooded massacre of unarmed civilians, the exact number of which was not yet known.

It was very early in the morning in Kandahar City when local journalist Mamoon Durrani was awakened by a telephone call. A war reporter for BBC Pashto, Agence France-Presse, and others, Durrani was being called by a contact he had in the local police department.

"He told me come quickly, something terrible has happened."

Durrani packed up his camera and his voice recorder and hurried out of his apartment and into his car. He drove west, toward Zangabad. Reaching the district where this terrible thing had happened, he was stopped by police.

"They told me no, you can't go any further. It's not safe."

But Durrani had a contact inside the Taliban. He called this man and received permission to pass into Zangabad.

As Durrani drove down the bomb-cratered road, headed toward

VSP Belambai, a new development was unfolding inside the outpost there. At 7:00 A.M., the officer working the night shift in the TOC learned more terrible news. A distraught Afghan man had come to the front gate wanting to speak with an interpreter, "a guy that lived about 100 meters off of VSP Belambai who said he had four dead bodies in his house." The man also said that "in another village, a whole family had been killed."

At Kandahar Airfield, at 8:05 A.M., senior army officers gathered for an emergency meeting to discuss what was being called "Consequence Management" in official paperwork. Outside the gate at the outpost in Belambai Village, two hundred local Afghan men had gathered, demanding justice. The situation was growing increasingly hostile and complex. U.S. officials in Kabul and Kandahar feared another riot like what happened after the Koran burning. Around Belambai, dispute resolution was handled by pro-Taliban village elders. In the aftermath of the massacre, the villagers wanted tribal justice, not a Western-style rule of law they were wholly unfamiliar with. The crowd demanded the Americans hand over the soldier who had committed these murders. They wanted him hanged.

Inside the fortified garrison, the Green Berets prepared for a riot. "We had a team that was ready to go," Captain Fields told investigators. "They had their riot-control gear [on] because the Quran burning was two weeks prior to this." Down the road, an army officer from FOB Masum Ghar was charged with handling the volatile situation outside the VSP. Colonel Todd R. Wood, commander of the 1st Stryker Brigade Combat Team, 25th Infantry Division, would run point on Consequence Management. To "get out in front of [the violence]" that was unfolding, he said.

As the news came in, the scope of the horror was far greater than initially understood. Bales had shot and killed sixteen people in three separate locations. Nine of them were children. He'd wounded more than a dozen other people. Had tried to burn bodies in two separate fire pits. At least one cow and one dog had been shot, killed, and set on fire. There were three major crime scenes that needed to be investigated. In a working criminal justice system, a crime of this magnitude would be sealed off and combed over for forensic evidence. But there were no local police. The area was too dangerous for American criminal investigators. There would be no crime scene investigators for twenty-two days. Journalist Mamoon Durrani was the only non-villager at the scene. He was shown the bodies and allowed to take photographs, which were later reprinted around the world. The scene was so horrific, he momentarily lost his professional composure.

"One of the little girls [killed] was so small I began to cry," he recalled in 2019. "The villagers could not understand why I was crying. 'You are not from around here, why are you crying for us?' they kept asking me."

Around 11:20 A.M., a four-vehicle convoy of armored Mine Resistant Ambush Protected All-Terrain Vehicles, or MATVs, arrived in the village carrying Colonel Wood and a small security team. The colonel got out and made his way into the crowd. "As I got closer, I made eye contact with [Tribal Affairs Minister] Haji Khudaidad," he remembered. "He motioned me over, he wanted to talk to me, so I moved then by the vehicles, and at the first vehicle there was an Afghan Army soldier who recognized me."

Colonel Wood observed four open-end trucks, "what we would refer to as a bongo truck," he remembered, "sort of [a] modified pickup used to haul anything in." The village elder waved him closer. People who had lined up along the dirt road stepped aside so the American colonel could see. One of the villagers pulled the tarp from the bodies. In one truck after the next, Wood recalled observing bodies piled up like cordwood. Burned corpses. Charred limbs. Clothing soaked in blood. Wood nodded to the tribal leader, indicating the funeral procession was free to go. The convoy headed to the cemetery, where the murder victims were buried, their graves covered in a tall pile of rocks.

In a strange kind of symmetry, as the bodies were being examined by Colonel Wood, on the back side of VSP Belambai a U.S. Army Black Hawk helicopter touched down, and one CID special agent and two Camp Brown 7th Special Forces Group staff sergeants climbed out. They secured Staff Sgt. Robert Bales with handcuffs and prepared to take him to Kandahar Airfield, making sure Bales had his body armor on over his uniform, in case he was shot at by the crowd.

Colonel Wood reported back to the Defense Department's Consequence Management team, in Kabul, to let them know what the situation was like on the ground. News of the Kandahar Massacre had by now spread like wildfire in newspapers around the world. The U.S. team determined the environment was too hostile for American criminal investigators and law enforcement officers to get anywhere near the multiple crime scenes. Instead, the chief of police from Kandahar, a notoriously corrupt official named Brigadier General Raziq, dispatched a unit of Afghan Uniformed Police to go to the villages and collect evidence the following day.

On the morning of March 12, Raziq's men went to the murder scenes. Their investigation "lasted one hour." They turned over to U.S. Army CID agents a total of eight shell casings—five spent rifle shell casings and three spent pistol shell casings—that they found at three major crime scenes where it was initially believed seventeen people had been killed. Why so few shell casings? "I'll give you a couple of examples [why]," one of the American soldiers at VSP Belambai told investigators. "Whenever we'd get into firefights, there would be ANA soldiers with empty [ammunition] mags. Others running around trying to pick up brass, because they would sell the brass at the bazaar."

The official Afghan Uniformed Police report, translated from Dari, listed the "Identities of those killed" as:

- Mohammad Dawood son of [name redacted]
- Khudaidad son of [name redacted]

- Nazar Mohammad son of [name redacted]
- Payendo's daughter (adult)
- Robina daughter of (unknown) 6 years old
- 1 unknown guest in [name redacted] household
- 11 members of [name redacted] family that their identities are unknown, the majority of which were children and were killed

Twelve of the seventeen victims were nameless people. Meaning, their identities were unknown to the U.S. military as well as the Afghan police. That no one could determine who the individuals were added to the tragedy of it all. There was more misinformation produced when it was determined that General Raziq's police force counted the number of murder victims incorrectly. Later the dead would be reduced in number from seventeen to sixteen. If the goal of the Afghan 1000 program was "to enable the government of Afghanistan to more clearly identify who its citizens are," the Bales massacre made clear this was a hopeless fantasy. Implementing the rule of law in Taliban terrain was like trying to light a match in a hurricane.

In Kabul, President Hamid Karzai wanted to send a delegation of his own to investigate, but this would take another three days. As Karzai's Afghan National Army chief of staff, General Sher Mohammad Karimi, and his deputies made their way through the village, they came under direct fire. A Taliban sniper shot and killed one of the Afghan soldiers guarding Karzai's top general.

In lieu of an immediate criminal investigation, and for the purposes of Consequence Management, it was decided by the U.S. Army chief of staff that a million-dollar solatia, or condolence, payment to the families was the next best move. "Money is my most important ammunition in this war," then Major General David Petraeus wrote in the army's *Commander's Guide to Money as a Weapons System.*

Colonel Wood briefed his commander about why he believed

this would put an end to the dispute. "In summary what I would share with you is the KLE [Key Leader Engagement] management [team] and consequence management [team] was [*sic*] aggressive and it was based on experience," Wood said. "We've learned from the Koran burning, how to get out in front of it; and I would say we did get out in front of it."

Staff Sgt. Robert Bales was whisked out of the country. Afghan tribal leaders convinced the victims' family members to take money as an acceptable means of dispute resolution. "Because of the condolence payments, the families are moving on," Colonel Wood wrote in an official report.

Three days later, on March 16, one of the victims' family members, Habibullah Naim, traveled to Kabul to meet with President Karzai. Naim was the villager on the motorcycle who helped save the lives of the five shooting victims that first night. With pomp and ceremony, Naim delivered his eyewitness summary to President Karzai and a delegation of men. The event was recorded and broadcast for all Afghans to see. As the delegates sat and listened, they sipped tea from white china cups.

On March 23, 2012, the U.S. Army formally charged Staff Sgt. Robert Bales with seventeen counts of murder and six counts of attempted murder, for a total of twenty-three victims, all of whose names were redacted. The army's incorrect count of murder victims was later corrected to sixteen. The U.S. Army CID had yet to investigate the crime scene, instead deferring to a poorly documented "official" Afghan version of events.

The next day, solatia payments totaling $970,000 were made to the victims' families. Colonel Wood oversaw the handover of money. "We developed through various source documents a list of possible claimants," Wood later told investigators. "Eventually accepting" that there were four men who would be paid on behalf of what were now twenty-two victims in total. "We brought out the money," Wood recalled, "and placed it in a box, and gave them a few plastic bags . . . At the end of paying everyone, we sealed each box so that their names were [written] on the boxes." Kandahar police chief Brigadier General Raziq escorted the families away. "The U.S. had spoken about taking them to a local bank in Kandahar, but that did not happen," Wood said. No one from the army's Criminal Investigation Command had yet been to the crime scene.

The tragedy did not end there. After Habibullah Naim flew to Kandahar to meet with President Karzai, he went back to farming the land near VSP Belambai. After the Kandahar Massacre, the Taliban gained a stronger foothold in the local villages and continued to sow the terrain with IEDs.

"Habibullah needed to farm his land," Mamoon Durrani explained to me in 2019. "So he asked the Taliban how to disconnect the wires on the chain of IEDs." This was not an uncommon event, as PGSS operators interviewed for this book confirmed. "To the U.S. forces," watching from overhead surveillance platforms, "this action made him an insurgent," Durrani says. Habibullah Naim was targeted and killed by U.S. forces in an air strike, on his own farmland, on October 5, 2012. "He was the only brave witness of the Kandahar Massacre," Durrani says. Now he, too, was dead.

What this one American soldier did was unconscionable, Durrani says, referring to Bales. It was also "one man, one soldier with a [malevolent] heart." What the Taliban does, and will continue to do now that they have regained power, is far worse, Durrani fears. In 2020, he sent me a photograph he had recently taken at the Kandahar police station. The picture was of a corpse he'd been asked to try to identify. The body had been so disfigured I had to ask Durrani what I was looking at. "The Taliban has started boiling people alive," he said.

CHAPTER EIGHT

BATTLE DAMAGE ASSESSMENT

A few days after the Kandahar Massacre, First Platoon was sent to a highly contested area located eleven miles to the northwest of Belambai Village, where Staff Sergeant Bales had just killed sixteen people. Their new outpost was called Check Point AJ. No one knew what the "AJ" stood for.

"Most likely A and J were the letters of the name of a local village," Private Twist guessed. The checkpoint was way out in terrain the U.S. military failed to control.

"The whole area down there was what we called a biometric hot spot," 1st Lieutenant Grant Elliot, the assistant S2 (intelligence) for the squadron, explained. "Many enemy fighters previously identified as Taliban used this area as a bed-down location."

The S2 shop, located in an air-conditioned tactical operations center at Forward Operating Base Sarkari Karez, wanted as much biometric data about people living in this contested area as it could get. The geography around Check Point AJ was laden with an ancient karez irrigation system, Elliot recalled, a network of subterranean tunnels in which the Taliban stored weapons caches. These weapons contained fingerprints and DNA—data that could link people to events and help S2 create what were called link analysis charts to build out intelligence on IED cell networks for future use.

"Our goal for a biometric hot spot," said Elliot, "was to locate weapons . . . and enroll a lot of people into BAT/HIIDE [devices] to identify potential targets on the watch list." First Platoon's lower enlisted paratroopers were not privy to S2 intelligence efforts. Most had no idea about the role of biometrics and had never heard of a link analysis chart.

The army's relationships with locals was strained before the Bales massacre. Now it felt untenable, Staff Sergeant Williams recalled. When the platoon first arrived, Williams was given three warnings by the noncommissioned officer he replaced.

—Never go out at night.
—Never go out without an armored vehicle.
—Never go more than 200 meters east.

It did not take long for the soldiers to realize the checkpoint was shoddily built. While Sergeant Williams was talking with another NCO, Private Twist lit up a cigarette to enjoy in the hot winter sun. As Twist was leaning against the wire mesh wall of a HESCO barrier, it nearly collapsed.

"They were only half-filled with dirt," remembered Private Walley.

Partially filled HESCOs were dangerous. "They made a perimeter easier to breach," remembered Williams. No one wanted to be out building a fort in Taliban territory, but this needed to be done—and fast.

"Standing around with your head down shoveling sand" made you a target for an insurgent with an AK, Sergeant Herrmann said.

First Platoon's paratroopers formed a bucket brigade, passing bags of sand down a line. Most of the Afghan National Army soldiers sat around watching, except one soldier who volunteered to help.

"He made a game out of it," Private Thomas remembered. "He'd throw the bag of sand to me, I'd toss it to the next guy, who tossed it to the next guy." This gave Thomas an optimistic feeling about partnership with the Afghans.

"That, or get paranoid about green-on-blue attacks," said Twist.

"I don't remember his name, but he became my friend," Thomas said. "He didn't have to help us that day, but he did."

Age eighteen, Pfc. Zachary Thomas was the youngest soldier in the platoon. Growing up in Texas, he hunted deer with his father and made sausage from the meat.

"I was an entrepreneur as a kid," Thomas said. "Got up early before school and made sausage sandwiches, which I sold." Everyone bought them, students, teachers, the bus driver. "Two dollars apiece." Thomas knew how to laser focus on one thing. In high school he won competitions for orienteering, a form of long-distance racing with a compass and a map. "I wasn't good at long-term goals like studying, but I pushed myself to win whatever was right in front of me. Pain is in the head, not the body, I'd tell myself." This mindset helped him in Afghanistan, he said. If you lost focus, you could die.

In these early days of the deployment, Thomas saw the experience as something he could learn from. Not just the soldiering part of it, but in the strange and foreign nature of the people, the culture, the terrain. Once, on patrol, he found an Afghan hedgehog. Reddish colored and with long ears. Another time he found a frog, identical in color to the camouflage uniforms they all wore. Keeping guard in the tower at Check Point AJ one night, he noticed the ANA soldier he was working alongside had six fingers on one of his hands.

"I remember thinking it had to have been hard for him to shoot," Thomas said. The two soldiers did not speak the same language, so Thomas conveyed a suggestion, using his hands to make gestures.

"I told him a doctor could remove it. Then it'd be easier for him to fire his weapon." But the Afghan soldier did not like Thomas's suggestion, he recalled. "He got angry. Kept pointing at the sky and saying God made him that way. Allah. He was not gonna have his finger removed. No way."

Thomas changed the subject, and the two soldiers kept guard in silence. The night skies in Afghanistan were majestic. Black and filled with stars.

Shortly after the platoon got to Check Point AJ, Sergeant Williams realized the platoon was being used for something other than their idea of what soldiering was.

"It was a shitty feeling," said Twist, one that was "easier to ignore" than confront.

"It was messed up," said Specialist Haggard.

"All those military tactics we learned at Bragg?" remembered Twist. "Never mind. That training to walk in a wedge? Not anymore. It's dismounted foot patrol. Single-file line."

"All they wanted was [for] us to agitate Taliban," Sergeant Williams came to realize. "Draw fire. Cause a firefight . . . and do a BDA."

"All they cared about was the damn BDA," Walley said.

A battle damage assessment, or BDA, assesses the amount of damage inflicted on a battlefield target. The origins of the BDA, during World War II, were in infantry troops sent to gather data from an aerial bombardment campaign. The idea was to bring back damage data for officers and analysts to evaluate. To determine if the bombardment was effective or not. Twenty-first-century biometrics had moved the concept of BDA into new territory, to include damage assessment of dead humans.

Every time a fighter got killed in Afghanistan, the Company Intelligence Support Team, or COIST, member was ordered to do a BDA.

"I conduct BDA," Pfc. James Skelton later told a military judge. "I was the COIST member attached to the Headquarters element of [the platoon] . . . I'd do intel analysis, BDA, intel gathering, biometric entries . . . I carried a SEEK biometric-entry device that I could enroll local nationals into, and we can check to see if they have any history with coalition forces, good or bad."

Getting BDAs and gathering biometrics. That was what the army was after in 2012.

The job of COIST did not exist until 2007. That is when a hard-charging army colonel named Harry D. Tunnell IV developed the concept. Since the position did not previously exist, there was no official training program. There were tactics, techniques, and procedures to learn, which took place at the National Training Center at Fort Irwin, California, and Fort A. P. Hill in Virginia. Private Skelton and the other COIST members went there to train on how to use the BAT, the HIIDE, and the SEEK. To write storyboards after foot patrols. Prepare reports for the company commander.

Soldiers chosen for the job of COIST were recycled from jobs that were no longer needed in the counterinsurgency fight. "[A] soldier, for example, is assigned as the chemical operations NCO . . . He really doesn't have a job in the current global war on terror, because there's not a chemical fight," Captain Edward Graham told an army reporter. This person would be an ideal choice for a COIST member. "We [give] him all the tools that he need[s] to be able to take the current battlefield picture, understand its dynamics . . . and be able to put all that information together and develop further on targets," said Graham. "Now he can be focused on being an analyst." The job was not coveted. From company commander on down, biometrics and BDA were perceived as a bureaucratic nuisance.

"Biometrics was that thing you were required to deal with,"

Captain Swanson said. "But no one really wanted to be dealing with it."

And yet higher up the chain of command, the opposite was true. The job of COIST, and the information the COIST member provided, was crucial for battalion- and brigade-level commanders. In January 2010, when the army issued its first COIST Handbook, Brigadier General James Yarbrough cautioned how the job of COIST, and the biometric data the COIST was responsible for collecting, upended centuries of intelligence reporting in the chain of command. COIST created what General Yarbrough called "Bottom-up Intelligence." Small units operating on the ground would now be the ones to "gather and determine the significance of intelligence, often without the assistance, analysis, and filtering of higher-level staff support," the general warned. COIST members would actually have "brief periods of situational understanding and information superiority" over commanders far up the chain of command.

This created friction within the platoon. Unaware of the role that biometrics collection played in the bigger context, Private Skelton was looked down upon by many of the other privates. Combat infantry platoons place value on paratrooper wings, weapons prowess, PT (physical training), valor on the battlefield. Skelton was a former traffic cop from Southern Pines, North Carolina. That he was perceived as overweight added to the other problems. First Sergeant Kenneth Franco openly made fun of Skelton.

"Called him disgusting," remembered Walley.

"First Sergeant Franco forbid Skelton from wearing his tight-fitting shirt," remembered Williams.

In a homogenous group, and where everyone is encouraged to work together, problems arise when one person is ostracized.

"Skelton only looked out for Skelton," Williams said. "And only Skelton."

In retrospect, Lieutenant Elliot blamed the discord on bureaucracy. On the way the military chain of command was structured and laid out in Afghanistan. "There certainly was a distrust of the purpose of the COIST [member] and what his role was," Elliot recollected in 2020. "At the company level [that is, the soldiers out on patrol], they were *reacting* to enemy behavior. And at the squadron level in the intelligence shop [that is, for the soldiers in the tactical operations center], we were trying to *predict* enemy behavior." The COIST had to act as liaison between these two disparate groups. Between biometric hot spots in highly contested areas, and the air-conditioned operations centers in the larger, more secure bases where the S2 analysts worked.

The disparity between these two realities came to a head during a dismounted foot patrol during the first week in April 2012. Private Walley remembers leaving the checkpoint in the middle of the night. Specialist Fitzgerald remembers being briefed on a BOLO, to be on the lookout for someone wearing a red cap. "An older guy. That's all we were told. Go on this mission and locate this man in the red hat." Private Thomas remembers being under the impression the patrol would be brief. "We were supposed to walk a hundred meters out, up a hill, check on a compound or something, and come back." Sergeant Williams remembers the mission being about a meeting with a village elder. "We were supposed to go find this guy and have a *shura*," a meeting, "with him at his compound."

The patrol reached the village around sunup. Everything was the color of dirt. The mud-brick houses, the dirt roads, the mud walls pockmarked with bullet holes. A goat or a cow here or there. Plastic trash to avoid stepping on at all costs. Some trees with no leaves, like stick figures stuck in sand. Marijuana and opium-poppy fields. To the southeast, a cluster of homes along the Arghandab River. Beyond that, more sand. The Registan Desert stretching all the way to Pakistan.

The platoon's dog handler and the bomb dog were out front. The sun was beating down. It was hot, the dog was overworked and would soon lose its ability to smell.

"Dogs were always losing their ability to smell," recalled Fitz. "Same as [the] Minehound was always running out of batteries."

Private 1st Class Matthew Hanes, as minesweeper, was right behind the dog, swinging the long arm of the detection unit out in front.

"Sweep or die!" Hanes would cheer, his dark humor beating back the ominous presence of IEDs. Foot patrols felt like death marches if you thought about it too hard. Herrmann's booming "Pay attention or die" undergirded every step. These patrols were the brainchild of Gen. David Petraeus, commander of international forces in Afghanistan from 2010 to 2011. "Walk. Stop by [villages], don't drive by," Petraeus wrote in a 2010 guidance memo for 100,000 soldiers, sailors, airmen, marines, and civilians to obey. "Patrol on foot wherever possible and engage the population. Take off your sunglasses . . . [Do] not [be] separated by ballistic glass or Oakleys." David Petraeus was CIA director now, and here was First Platoon.

Hanes was from Philadelphia. His charm and good looks earned him the nickname Justin Bieber. Back at Fort Bragg, thrilled by earning money thanks to Uncle Sam, Hanes bought a car with his parents' help. "A bright, shiny, yellow Corvette," remembered Twist. Before they deployed, the two of them drove around base looking for girls.

Walking a few soldiers behind Hanes was Fitz, assigned to the squad that was out in front that day. "We saw a guy in a red cap," he remembered. "We followed him up the hill. Stopped him. He was a chubby teenager, not the fighter we were looking for. [The kid] being chubby, that was strange for Afghanistan."

The patrol continued on, the soldiers snaking along in a line behind Private Hanes. Finally, they located the compound where the meeting with the village elder was to take place.

"He was old," Sergeant Williams remembered. "He said he was a mujahideen trained by the Americans from the anti-Soviet days." Who knew if it was true. "Didn't matter. He invited us in."

"He made us tea and foot bread," remembered Private Thomas. "We called it foot bread because he kneaded the dough with his feet before he cooked it."

Lunch was interrupted by an order over the radio. It was Captain Swanson at the tactical operations center.

"PGSS [surveillance] camera spotted enemy PKMs [machine guns] on the rooftop of a compound [nearby]," Williams recalled. "We were supposed to go out into a field and draw fire so headquarters could use mortars on it . . . They wanted a BDA of the compound, but they had to destroy it first."

The soldiers began moving toward the compound identified as having the machine guns on the roof. As Williams got closer, he realized "the reporting was messed up. There were brooms on the roof, not machine guns."

Too late. The platoon was being shadowed by insurgents.

Private Thomas, the radio telephone operator and fire support team specialist, relayed the information as it came in.

"Get down! Get down!" he heard coming in over his radio. "Get down immediately!"

At the operations center, Captain Swanson and the headquarters element were watching a live feed from the persistent surveillance cameras. The platoon was surrounded by insurgents on three sides.

"Get down, take cover" spread across the platoon.

There was a wadi ahead. An irrigation ditch filled with water connecting to the karez. A place to take cover on the other side. But the only way across was by walking over a red pipe, like a bridge. Walley became agitated when one of the ANA soldiers cut in front of him.

"I was pissed," remembered Walley. Then the sound of a bullet. Then a visual right in front of him. A bullet struck the Afghan soldier in the head. The man fell back into the wadi. "He was where I would have been," Walley remembered. "Just like that," life was over for him. Face-up, back down. Dead. The soldiers opened fire, getting low to the ground and scrambling for cover. There was direct fire coming at them from three sides.

Sergeant Williams and Lieutenant Latino found a farmer who let the platoon shelter inside his compound. There were children inside. Walley remembered there were three or maybe four of them.

"The kids acted like they'd never seen American soldiers before. They wanted to know what we do." Someone had paper and a pen, and Walley began drawing pictures of paratroopers jumping out of airplanes. "Like a cartoon." The kids were enthralled. Then the ANA soldiers came running in.

"They were really upset," Thomas remembered. "One of them said, 'You left one of our guys down there.' And they were right. We left the ANA in the wadi because we were taking fire."

"He's dead," Sergeant Williams told the soldier.

"He's still alive," another one of the Afghan soldiers said.

First Platoon's medic was new to the unit. Fitz thought he hadn't been in a firefight before. "The doc said, 'If you get him here, I'll do my best to keep him alive.'"

Walley volunteered to go with some of the other soldiers, to get the ANA who'd been shot in the head. After a while, the soldiers came back with the man's body.

"He was dead," Walley remembered.

"But our medic was convinced he was alive," Williams recalled. "They laid him down on the floor of the farmer's house and his body moved a little. This happens. It's called the death rattle."

Private Thomas realized it was the Afghan soldier who had helped fill the HESCOs with sand. His friend.

"I had his head in my lap," Thomas recalled. "The doc undid his helmet and his brains came out in my hands."

Radio calls were coming in from headquarters, and Thomas had a job to do. Close air support, known as CAS, was coming in to take out the insurgents' compound by air assault. First, Apache attack helicopters. Then an aircraft would drop a 500-pound Joint Direct Attack Munition, known as a JDAM.

"Watching the air assault . . . it was the most damage I've ever seen an Apache do," Walley remembered of that day. "There's a lack of language to express this memory. It is the most terrifying thing you will ever see, but also the most satisfying thing because the people who are inflicting the damage are on your side." Walley paused to gather his thoughts, relaying the story in 2019. "That something exists that can do that much damage, you can't imagine anything like it until you see it for yourself."

He remembers how a U.S. aircraft dropped a 500-pound bomb approximately 100 meters away from where he was standing. "They call it danger close. A bomb like that takes the oxygen right out of your body. There's a vacuum all around you. Shrapnel fills the air. Then it's over and you're just lying there. I remember looking up at the sky and seeing a plane way, way up above. Like an ant in the sky."

Private 1st Class Nicholas Carson was with Walley and the two of them climbed up on a structure near the farmer's compound, to fire the Carl Gustaf rifle at the Taliban from up there on the roof. Walley provided cover, and he remembers thinking how odd it all was. "A straw shed with some goats. There was a blown-up car. A few other animals roaming around. Someone had written something on a wall, like graffiti," in a language he didn't understand. Then, back to the farmer's compound, where Lieutenant Latino and Sergeant Williams were arguing with headquarters over the radio.

"We can't do a BDA, we're still in a firefight!" Lieutenant Latino shouted.

"Trying to do a BDA is impossible," Sergeant Williams said.

"Word came back over the radio. 'We're doing the BDA,'" Thomas confirmed.

Williams and Latino tried stalling. If they could wait until darkness, their night-vision goggles would give them a technological edge.

"But headquarters wouldn't let up," said Williams. "All they wanted was that damn BDA."

Soldiers follow orders; this is the way of war. The platoon broke up into squads and made several attempts at the BDA this way. The first attempt at reaching the compound failed. A group of soldiers got halfway across the field with the bomb dog when the Taliban opened up on them, pinning them down.

One attempt after the next failed. Only when darkness fell was the BDA achieved. Sergeant Williams and Lieutenant Latino led a group of soldiers to one part of the destroyed compound, through an irrigation ditch filled with snakes and frogs. A second group pulled security while Staff Sergeant Nuanes, as EOD tech, and Private Skelton, as COIST, did their jobs, Fitzgerald recalled. Fingerprints and DNA needed to be gathered, bagged and tagged, and sent to the lab in Kandahar to be analyzed.

"Iris scans from dead Taliban? Half of them didn't have any heads left," Walley recalled.

Fitzgerald remembered most of the bodies inside the compound as being "charred beyond recognition." And yet, "Skelton did his thing. Everyone was dead. I hated how it smelled."

Walking back to an army outpost or checkpoint in the dead of night is rarely a good thing. In the rush to return, First Platoon got split up.

"The NCOs were taking everyone back through the river," Walley recalled. "I was standing near Nuanes, who said, 'Follow me.'"

"The EOD always knows best," recalled Twist, "about everything."

Walley and Private David Shilo, on heavy weapons, followed Nuanes. They made it back to Check Point AJ walking along the edge of a tributary. The rest of the platoon exfiltrated back to base through the wadi.

"When we finally got back," Walley remembered, "we were dry; they were soaking wet. Damn if the EOD doesn't always know best."

The time at Check Point AJ lasted five weeks. Just as they were settling into a battle rhythm, an unexpected change of plans. Time to move out, Captain Swanson informed the troop.

"Soldiers with the Second Infantry Division had fired a grenade into the tree line and killed a kid," Swanson revealed in an interview for this book. "And maybe a mother too." The killings happened approximately nineteen miles to the east, in a remote and contested area in Zhari District called Payenzai.

"Colonel [Scott] Halsted had gone there with a lot of money to try and make things right," for the family of the civilians killed, Swanson clarified. This was to be a solatia payment, compensatory money, same as had been paid to the grieving family members related to those killed by Staff Sergeant Bales.

President Karzai had promised the nation there would be no more civilian casualties, and now there was yet another one. General Allen had gone on television and given his word.

"The platoon that was there," Captain Swanson explained, "with Second Infantry Division. They needed to get out of there after what happened, and we were going to Payenzai to take over."

Word made its way down the platoon: "Second ID killed a kid."

"We were supposed to keep the villagers from getting angry," Walley recalled.

"To keep some kind of riot from breaking out," Twist remembered.

How would the villagers know the difference between one set of U.S. Army soldiers and another? Everyone wore the same camouflage uniforms. Wore the same helmets and body armor. Carried similar weapons.

"In southern Afghanistan, people know other people as tribes," Walley explained. "Second ID was [known] to the villagers as the Indian Head Tribe because they wore a patch [on the arm] with an Indian Head. The Special Forces dudes? They were the Bearded Tribe. We were the Circle Square Tribe because of our patch, two A's for All American, that looked like a circle inside a square."

"We were different from the Indian Head Tribe," remembered Twist. "We were the Circle Square Tribe, and we [were] there to keep the villagers safe from the Taliban Tribe."

On May 12, 2012, the platoon was preparing to move to the new outpost, Strong Point Payenzai. EOD specialist Staff Sergeant Nuanes was needed down by the Arghandab River to accompany another platoon on a mission there. As First Platoon's soldiers were packing up their gear and getting reading to leave Check Point AJ for good, Sergeant Herrmann broke the terrible news.

"Nuanes is dead," he said.

"Nuanes bent over to examine an IED and the device blew up in his face," Twist recalled.

Walley could not believe what he'd just heard. "You soak in the emotion for about thirty minutes and then you go numb. I took it especially hard. Nuanes was insanely intelligent. Brave as shit. I had a level of respect for him that doesn't compare. Now he was dead."

The explosive ordnance disposal specialist knows best. Who would work with them now? And would this new EOD be as qualified and as patient as Israel Nuanes had been?

But there would be no EOD specialist assigned to First Platoon after Nuanes was killed. Foot patrols in Zhari District around Strong Point Payenzai would be like putting a blindfold on and then walking through a minefield. The worst was about to begin.

PART III

CHAPTER NINE

STRONG POINT PAYENZAI

Strong Point Payenzai was too small to be called a combat outpost or even a checkpoint. Built in a triangle shape, the garrison was constructed of HESCO barriers and topped with coils of concertina wire. There was no electricity, no phone line, no water. The soldiers cleaned themselves with baby wipes. When First Platoon arrived, there was no place to sleep. No army-issued Alaska tents, so they slept on wooden pallets left over from a provisions transport, stretched out under an endless constellation of stars.

"Staring up at the sky in Afghanistan took your breath away," remembered Spc. Anthony Reynoso. "You could almost forget there was a war." Everyone had a nickname. Reynoso's was Rey-Rey. No one could surpass him at physical training.

Each corner of the strongpoint had a guard tower constructed of plywood and sandbags. From here forward there would always be at least one soldier positioned inside each guard tower, scanning the tree line for an indication of ambush or attack. Two of the three towers were manned by American soldiers, the third was for the Afghan National Army. Strong Point Payenzai was one of more than 700 military garrisons constructed by the U.S. Army across Afghanistan. Here in the south, the intention was to have these forts work like a network of power spread out across the terrain.

First Platoon arrived at Strong Point Payenzai on foot—a grueling, eight-hour march from Strong Point Ghariban, which was one klick, or kilometer, to the northwest. For most of the privates, Payenzai would be a final destination. Ghariban was where company commander Captain Swanson would remain stationed with the Headquarters Unit and Second Platoon. In sum, they made up Chainsaw Troop, roughly 100 men. The soldiers would sometimes shuttle back and forth between the two garrisons in armored vehicles. Ghariban was positioned closer to the American-built Highway One, which made it closer to the supply line, more fortified, and more secure. Ghariban was built around one of the local villager compounds, and some of its mud-brick structures had been subsumed inside the walls of the army's new fort. There was a tree growing in the courtyard. No one realized it was an apple tree until one day it produced a lone fruit, eerie in its singularity and biblical connotations about knowledge and evil.

Ghariban was where the post-collection biometric efforts took place. Where Private Skelton, as COIST, would write up storyboards and biographical summaries of persons of interest encounters. There was electricity and a satellite hookup at Ghariban, and a secure area for biometrics. Only those with secret clearance could enter. The rest of the platoon did not have a need to know what went on there, and remained in the dark.

The idea of the strongpoint is tied to an age-old military concept of fortification. In *Winning Insurgent War,* Geoff Demarest addressed the effectiveness of fortification in southern Afghanistan in the form of a question, asking 82nd Airborne officers to consider how using forts shaped the counterinsurgency battle. Is it a "Maginot Line or

Fort Apache?" he asked. The French-built Maginot Line, infamous in World War II, was a line of concrete bunkers, fortifications, and barriers built by the French government in the 1930s to keep the Germans from invading again. The Maginot Line was supposed to be impenetrable and included soldier amenities like air-conditioned dining halls. The Nazis found its weakness and plowed through. What happened next is history.

Fort Apache, in Arizona, was constructed in 1870 and became part of the U.S. military campaign against the Apache Indians who'd been living in the region for centuries. Because Apache chief Geronimo surrendered to U.S. cavalry in 1886, Fort Apache is often portrayed as a success story for forts. But as the military historian T. Miller Maguire wrote in 1899, "a fortress once invested is certain to fail unless a relieving field-army can beat the besiegers away." During the Indian Wars, the U.S. Army was unwilling to give up even an inch of terrain taken from the native Indians on anything less than its own terms. This was hardly the case in Afghanistan. By 2011, the war was already being called a "hopeless failure" across Washington, DC.

"The government in Kabul, we have been told, is corrupt and predatory," wrote national security analyst Peter Bergen when explaining a phrase he helped bring to the public's attention. "The Afghan army is a mess. Tribal loyalties trump national loyalties. The Taliban is gaining in strength." The same month that most of First Platoon's privates were graduating from high school and preparing for basic training and airborne school, the concept of hopeless failure had become a talking point on the evening news.

Foot patrols and biometric collection began immediately at Strong Point Payenzai. Fingerprints and iris scans were to be taken of every villager the patrol came in contact with. The process always felt strained, remembered Private Twist. "It was supposed to be a hearts-and-minds campaign," he said. "Telling someone 'open your eyes wider . . . no, wider' again and again, this felt a little strange."

Spc. Dallas Haggard remembered the overreach too. Haggard was from Ohio, where he had been a high school football star. He had a dry sense of humor and a deep loyalty that guided him in his actions. If you got hurt, it was Haggard who would fireman-carry you to safety. "These people are farmers," he said in an interview in 2019. "Sun comes up, sun goes down. They need to make sure they take care of their grapes. Time is all they have. Asking a farmer to stop what he's doing, for an hour so, so we can grab some finger scans, some eye scans. It's like, what the fuck?"

Sergeant Herrmann found himself having to weigh decisions about possible consequences. "We say, 'Hey, we want eye scans and fingerprints.' They say no. There's the ethical question of how far should we push things?"

It was up to the officers, not the privates, to make the call. As a second lieutenant, Jared Meyer regularly found himself having to make this decision. "At what point do you use force? Say, 'All right, hands up, eyes open wide. Burkas off. Hands out of your pockets.'" It always felt like a quandary. "We were not trying to impose a police state."

There was an inherent conflict, said Twist. "We were supposed to let them know we're here to help."

With rule-of-law programs in Afghanistan guiding the mission, the armed forces' rules of engagement had changed.

"There were no longer night raids," remembered Staff Sergeant Joshua Giambelluca, who had participated in night raids in earlier deployments. "Soldiers could only fire their weapons if fired upon. It was like, fight with one hand tied behind your back."

"If we saw a villager talking on an ICOM radio and doing one other suspicious indicator like pointing or hiding, we could engage," Sergeant Herrmann recalled.

"In our little area of operation, which was less than, say, fifteen square miles, I estimate there were around one hundred military-aged males," remembered Captain Swanson, and that within the first few weeks of patrolling, "we'd pretty much captured biometrics on [everyone] who lived there."

Insurgents learn to be flexible. They observe the tactics, techniques, and procedures of the bigger army they are at war with, and adapt. To outmaneuver the U.S. Army in Afghanistan, as soon as night raids were prohibited, the Taliban did much of their work in the dark.

"We'd watch this activity at night, from our guard towers, through night-vision devices," Lieutenant Meyer explained. "We'd see foreigners come into our area at night and they'd leave by morning. We knew they were bringing in guns and bombs." Unable to take action at night, the platoons would have to wait until the morning to seek out information from villagers who were involved. "We'd go to them and ask them for information," he recalled. "They'd say, 'Ummm. I have no information.' Meaning 'Please don't ask us questions. If we tell you, something bad will happen to us.'" Lieutenant Meyer empathized with the villagers' situation. It was a Catch-22. "There were consequences for them, *and* for us. It was like they were saying, 'You don't want to die, but we don't want to die either.'"

The interminable nature of this dilemma was not lost on others.

"Later on," said Walley, "a big deal would be made over a villager with HME [homemade explosives] on his hands. If you walked around Payenzai and you swiped every villager's hand with Expray, damn near every one of them would test positive for HME." Expray is a product used by the army for detecting traces of homemade explosives on a person's hand.

"Put yourself in their shoes," suggested Private Twist. "You have to ask yourself, what would you do?"

Consider Habibullah Naim, just a few miles away in the village of Belambai. Killed in an air strike for turning off and on an IED belt set up by the Taliban on his farmland.

The biometric campaign on the ground had been designed to case-build rule-of-law arguments in the Afghan judiciary. To bring criminal cases against Taliban IED emplacers and Taliban bomb makers in Western-style courts of law. To use forensic evidence to discern between a farmer who was coerced into allowing Taliban IED activity on his land and a farmer who was an active insurgent. The way to do this, the U.S. Army believed, was by having machines comb through otherwise immeasurable amounts of data. It had been more than eight years since the Defense Science Board proposed to the secretary of defense the creation of a Manhattan Project–style biometrics program. Here in Afghanistan, this was that program in action.

But a new and unforeseen problem had arisen as a result. What the generals called a deluge of data.

Three years prior, as U.S. forces withdrew from Iraq, Air Force Deputy Chief of Staff Lieutenant General David Deptula warned colleagues of the coming data storm. "We're going to find ourselves in the not too distant future swimming in sensors and drowning in data," he said in 2009. In a single year, the air force had collected more video footage in Iraq than a person could watch twenty-four hours a day, seven days a week, over the course of twenty-four continuous years.

The problem was human. After years of surveillance video captured by drones, airships, and aircraft flying over Iraq, the U.S. Armed Forces found itself with millions of hours of video sitting in archives taking up space. Few people in the military knew what to do with all this data. How could they turn the Pentagon's vast digital archive into actionable intelligence? A privately owned American software company called Palantir Technologies proposed a solution.

Headquartered in Silicon Valley, Palantir was named after a fictional object in J.R.R. Tolkien's *The Lord of the Rings,* a crystal ball–like artifact that could see, and foresee, activities going on in other parts of the world. The company was cofounded by two entrepreneurs who met at Stanford Law School: Peter Thiel, whose previous venture was as cofounder of PayPal, and Alex Karp, a money manager. At Palantir, the two men created a software program that could sift through unstructured data, or raw data, then organize and structure that data in a way that made search and discovery features possible. This technique is shorthanded as "data mining."

"Well, we built Palantir starting in 2004 to supply the U.S. intelligence community with world-class software platforms, and we built one called PG [Palantir Gotham], which now powers a lot of the Western world," Alex Karp said in 2018, "that allows intelligence, defense, and police officials to more accurately track data, the provenance of data, and the people they are interested in tracking, in accordance with the law. And subsequently we built platforms for special operators in the more, kind of, data-aggregation space."

"Data aggregation" means turning data-mined, or organized, data into an intelligence product. "Space," in this case, means the war zone. The data packages, called "data slices" in military parlance, that Palantir produced now provided a solution to the storm of data coming in. Palantir could turn a flood of information into a single stream. Into actionable intelligence that the military and its intelligence community partners could legally take action on.

Palantir Technologies saw opportunities in partnering with the Defense Department in rule-of-law efforts around the world. The wars in Iraq and Afghanistan would likely be the first of countless future wars in the global war on terrorism. To garner a foothold in this digital domain would potentially pave the way for countless new business opportunities in both classified and civilian domains.

"The present and the future ability to control the rule of law and its application will be determined by our ability to harness and master Artificial Intelligence and its precursor, machine learning," Karp said. Personalized risk mitigation required an abundance of data collection, potentially on every human alive. Not in a lifetime could humans aggregate the data like Palantir's software could. It was engineered to search for, to locate, and to track individual persons. And to follow that individual's movements over space and across time. Who, or what, would assign a level of risk to each individual person was a decision that was still in the works.

As the Defense Department knew, archives of high-volume, high-velocity data were useless if analysts are not able to make connections between data points. Palantir's software programs made it possible to process, store, evaluate, and analyze data with a few simple keystrokes. Borrowing a metaphor from botany, consider an individual raspberry. Raspberries, in general, have many individual parts. Each raspberry is made up of canes, thorns, roots, leaves, and fruit, which are called drupelets. If someone were to take apart a million raspberries and mix up all the drupelets, Palantir's powerful software could, metaphorically, search through those millions of parts in order to locate and then aggregate, or collect, the precise and original drupelets of a singular, individual fruit.

"And so," said Karp, "Palantir went from being what you could see as a niche player for the intelligence community, the highly classified intelligence community [that is], to what you could call the platform upon which data analytics are done, inside and outside the U.S., in the course of ten years." By 2012, which was year eight in the ten-year timeline to which Karp refers, the U.S. military and its intelligence community partners in Afghanistan were using Palantir software to tag, track, and in many cases kill individual people it had identified as the insurgents who were building and planting IEDs.

•• •• ••

In the terrain around Strong Point Payenzai, the daily foot patrols were brutal. It was unimaginably hot. Wind blew pulverized dirt into every crevice of the body, every orifice, all day, every day. Sand blocked the nose and stung the eyes. A fine, gritty dust caked the fingers and never left the toes. Because IEDs had become omnipresent on the roads and goat paths in the area, the platoon was ordered to access local villages by traveling through farmers' land. The logic for this was that this terrain was less likely to be booby-trapped with IEDs. To cross a farmer's land meant climbing over five- to six-foot-tall walls called grape rows.

"It was a daily exercise in torture," remembered Private Twist. "Up, over, down. Repeat." Carrying sixty or more pounds of equipment in punishing heat. Always on alert for sniper fire from an insurgent hidden behind a wall, a grape row, or a tree.

In Maiwand District, around Check Point AJ, the Afghan farmers grew opium poppy and marijuana. Here in Zhari District, the main crop was grapes. Growing grapes in the desert would be impossible if not for the local invention of the grape row, these tall, mud-brick walls constructed row after row, and with trenches dug down into the water table below.

Specialist Fitzgerald remembered his initial impression. "When you first see the grape rows you think, How does that work? Then you realize it's ingenious. They dig down six, eight feet, where the ground is wet from the river. That's how they irrigate and make shade."

Zhari itself was a made-up district. In 2004, President Karzai created Zhari by carving it out of neighboring terrain, a reward to a local warlord who'd helped the Americans liberate Kandahar from the Taliban in late 2001. The following year, the Taliban returned, and Zhari District had remained ungovernable since. Canadian troops stationed in neighboring districts managed to reach the western edge of Zhari before turning their bases over to the U.S.

Army in 2010 and heading home. At the handover, commander Brigadier General Frederick "Ben" Hodges told reporters, "Zhari is probably about 90 per cent contested, or else flat-out under Taliban control, because we've never had the coalition or Afghan security force capacity to go out there."

When First Platoon arrived in Zhari District, the army had spent two long years trying to get a foothold in the area without success. One of the first big efforts, called Operation Dragon Strike, involved a series of high-profile kill-or-capture missions targeting Taliban commanders, but control remained elusive. In another failed effort, the army tried building a ten-mile-long concrete wall along Zhari District's southern border, destroying many farms and homes in the process. Villagers were offered financial compensation from the army's Civil Affairs Battalion, provided they filed a claim.

Notices went around the district. "[Come] to the claims center with proof of ownership of the property," said U.S. Army Major Benjamin Hastings. But less than 15 percent of the population could read, and property-ownership documents did not exist. District governor Niyaz Sarhadi intervened on behalf of the villagers. Title documents, he told the army, were a Western-world idea. A compromise was reached when the U.S. Army agreed that village elders could vouch for land ownership, so that villagers whose property had been ruined by the army could be paid.

But the Taliban kept coming, planting wave after wave of deadly IEDs. The army tried building a second concrete wall in yet another contested area of Zhari, this one bigger than the first. Task Force Strike installed hundreds of twenty-foot-tall concrete barriers along newly bulldozed roads, sometimes at the astounding pace of thirty to forty barriers a day. "Their biggest day so far is 68 barriers," 1st Lieutenant Allen Anders told a reporter in 2011. But the wall did nothing to keep the Taliban out.

By the time First Platoon arrived here in May 2012, the Taliban remained in control.

•• •• ••

At Strong Point Payenzai, day after day, every day, upward of ten hours a day were spent conducting dismounted presence patrols. Here were U.S. Army soldiers patrolling rural Afghan villages like cops on a beat. There was one foot patrol first thing in the morning, at sunup, and another one in the afternoon. The soldiers learned to navigate interactions with villagers; the goal was to make their strength and presence known. Encounters ran the gamut from friendly to hostile. But every interaction was a bargain. A quid pro quo.

"If you turn in IEDs, you get money," said Private Twist.

"If you get hurt by a soldier, you get money," said Captain Swanson.

"If you work for us as an informant, you get money," said Lieutenant Meyer.

"If you bring people to the weekly shura, you get money," said Sergeant Williams. "But before a villager got his money, he got fingerprinted, got his irises scanned. Got his cheek swabbed for DNA."

For the soldiers, one growing source of resentment around these foot patrols came from the fact that the NCOs carried with them GPS equipment to track and record their movements. Brigade headquarters required the data gathered get submitted after each patrol, a process known as honesty traces, or honesty trackers. Many of the privates felt like this requirement was an insult to their integrity. They were paratroopers with the 82nd Airborne, army elite.

"It felt like, 'Oh, we don't trust that you really went on that patrol,'" explained Twist.

These honesty traces were, in fact, part of something else data-related. An integral part of something much, much bigger. Something that, down the road, would leave the battlefield and become part of law enforcement efforts in the United States.

CHAPTER TEN

THE GOD'S-EYE VIEW

Two miles to the southwest of Strong Point Payenzai, down closer to the Arghandab River, inside a steel container set up inside Combat Outpost Siah Choy, Persistent Ground Surveillance System (PGSS) aerostat operator Kevin H was watching a man defecate in a farmer's field. The man was a person of interest, and Kevin H was working to establish his pattern of life in pursuit of a new intelligence-world methodology called activity-based intelligence. The first fundamental of activity-based intelligence presumes: *You are what you do.*

"Because of the high-powered equipment I was using, I could see everything," Kevin H says, referring to an omnipotence sometimes called the God's-Eye View. "The only way I didn't see something was if I wasn't looking at it."

Kevin H and the PGSS team spent much of their time watching First Platoon from overhead, but in this moment watching the person of interest going to the bathroom was a higher priority. There was a question at issue of whether he was squatting down to defecate or if he was squatting down to bury an IED.

The PGSS aerostat that Kevin H was in charge of was a seventy-two-foot-long balloon called a 22M (for meters) in defense-contractor parlance. It was not a dirigible, meaning it was not steerable and did not navigate through the air on its own power. The 22M was tethered to a mooring station inside the combat outpost at Siah Choy, attached by a 2,700-foot cable made of fiber optics, rubber, and Kevlar wrap. The flatbed surface on the mooring station reminded Kevin H of a merry-go-round because it could rotate 360 degrees unhindered. "It could swivel back and forth to allow for wind relief [which] mattered in the summer months, when the 120 Days of Wind kicked in." These powerful winds that would sweep across southern Afghanistan reminded Kevin H of the Santa Ana winds that behaved similarly across Southern California, where he grew up.

The equipment attached to the balloon afforded Kevin H a clear view of the soldiers, the strongpoint, and its environs. For the most part, the soldiers at Strong Point Payenzai were unaware that an aerostat with a suite of electro-optical-infrared high-definition sensors and cameras was able to watch them as they walked around their area of operations, through grape rows and into the villages on their twice-daily presence patrols.

"The idea was, do not let anyone know we exist," Kevin H clarifies. "Occasionally one of the Joes"—defense-contractor vernacular for soldiers—"an NCO usually, would use us as a way of saying 'We are watching you.' And we'd say, No, no, no, don't do that. We'd end up with some villager at our COP saying, 'I know you're watching us. Tell me who stole my goat.' This actually happened."

The MX-15 imaging system attached to the underbelly of the aerostat was roughly the size of a beach ball. It weighed 100 pounds and carried an array of cameras for intelligence, surveillance, and reconnaissance purposes. It could make out an unusual modification on the buttstock of an AK-47 from two miles away, Kevin H says. Because the balloon was lighter than air and got its lift through buoyant gas, it was difficult to shoot down. If an insurgent managed to put a bullet in its skin, the balloon could be reeled in and repaired.

The person of interest Kevin H was watching on this particular day wore a purple hat. Purple was an uncommon color for a hat in Zhari District, and so it became a unique marker for the man. The PGSS team at Siah Choy had been tracking him for weeks. Kevin H is a subject matter expert in pattern of life analysis, a discipline that involves establishing a person's identity based on their cumulative habits. The surveillance team had determined the man in the purple hat was a bomb emplacer, meaning he buried IEDs for the Taliban. The team had established his bed-down location. He lived across the river, on the south side. Like many of the people Kevin H and his team tracked, this individual insurgent was still an anonymous person to them.

"He would get up every morning, turn on an IED belt, a series of IEDs strung together by det [detonation] cord, to protect himself and his perimeter," Kevin H recalled. "We elevated him to 429 Status through his actions."

429 Status is what happens when a person of interest completes three "interactions with the ground." These are actions that allow for that individual to be moved out of civilian status and into insurgent status. To be targeted and killed legally according to army rules of engagement.

The three interactions with the ground were specific; Kevin H clarified by example. "If I see him interacting with the ground, and then I see the pressure tank going in, or [if] I see the charge going in, and him stringing the lamp cord out to install his pressure plate or his battery connection . . . That's one, two, and three for me." This is activity-based intelligence acquired through persistent surveillance from above.

As the team leader and mission director for aerostat operations at Siah Choy, Kevin H was in charge of a six-man surveillance team. Twenty-four hours a day, seven days a week, the team "covered down" on four infantry platoons patrolling a five-square-mile area that included twenty or so small villages. First Platoon and Second Platoon were two of the four platoons the PGSS operators watched. Earlier in the winter, the team's area of operations had been approximately twenty-five square miles, but by May things had gotten so kinetic, or lethal, their cover-down area had been significantly reduced in size. "There was an unusually high volume of activity going on" around Strong Point Payenzai and its environs, Kevin H recalled. A "constant flow" of insurgents planting IEDs.

At Siah Choy, the PGSS team worked inside a fourteen-foot command and control shelter, a shipping container used as an operations center. On one wall, two forty-inch monitors played image feeds from the MX-15. A third screen monitored the life of the balloon, a fourth screen monitored lightning strikes, which were dangerous to the functionality of the system. In the lights-off darkness of the shipping container, the PGSS operators watched the camera feeds incessantly. When a significant action occurred, they would relay that information over mIRC (military Internet relay chat) or telephone line to the 82nd Airborne's TOC at Siah Choy, located on the other side of the roughly 300-by-400-square-meter fortified outpost.

As a pattern of life expert, Kevin H spent twelve-hour days, seven days a week, watching villagers go about their daily lives with the goal of separating out the insurgents from the civilians. Pattern of life analysis means watching seemingly innocent behavior hour after hour, eyes focused for when the behavior of a person of interest might take a unique turn. For billions of humans around the world, one of the first actions of the day is defecation. In the Western world, this happens almost entirely in private, because most bathrooms in the Western world are located under a roof, behind walls, where an overhead surveillance camera cannot yet see. In southern

Afghanistan, there are very few toilets, which means almost everyone defecates outside.

"I have literally watched thousands and thousands of shits being taken in Afghanistan," Kevin H says, with a focus on what other interactions might, or might not, be happening while the person of interest is in the squatting-down position. "There are only a few reasons to squat in a field in Afghanistan," he clarifies. "To take a shit, farm your land, or bury an IED, which takes about the same time. Four minutes." The heat signature being registered by the aerostat's electro-optical-infrared cameras offer a pattern of life expert further clues. "Waste comes out of the body at 98.6 degrees; you can see that. So if that's what's happening, you're done watching." But if the squatter isn't defecating, then the surveillance at this point becomes about "How is this person interacting with the dirt? Do I see them connect wires? Do I see them covering [something] up with earth?" The accuracy of what an aerostat operator sees, and then reports, cannot be understated. "You don't want to fucking kill someone for taking a shit."

In 2012, while working at Combat Outpost Siah Choy, Kevin H was employed as a flight engineer for Navmar Applied Sciences Corporation, a defense contractor that provides "engineering and technical services" to support the Department of Defense and the intelligence community. In business since 1977, much of what the company does is classified but on occasion makes its way into the public domain. In 2017, *Wired* did a feature story on Navmar's ArcticShark, one of the company's unmanned aerial vehicles that conducts atmospheric tests in extreme cold-weather environments. The emphasis of this article was on the drone's ability to collect atmospheric data for climate-change analysis, an intelligence discipline called measurement and signature intelligence, or MASINT.

MASINT dates back to the Cold War and began in the 1950s with the CIA's U-2 spy plane as it searched to detect Soviet atomic-bomb tests by measuring radioactive particles in the air. With the advent of these high-flying aircraft, MASINT collection could scientifically prove a hypothetical the way a CIA spy could not, through human intelligence, or HUMINT. The U-2's onboard cameras further advanced the discipline of imagery intelligence, or IMINT, information gleaned from images. During this heyday of defense-science programs, the CIA's partner agencies advanced signals intelligence, or SIGINT, intelligence garnered from intercepted signals or communications.

The early U-2 overflights, and the A-12 Oxcart spy plane overflights that followed, played a major role in the advancement of the "INTs," the various intelligence disciplines that have become the centerpiece of national security in the modern era. Starting in the global war on terrorism, a new intelligence discipline evolved called geospatial intelligence, or GEOINT. It was James R. Clapper, a retired lieutenant general in the U.S. Air Force, who first officially unveiled the concept of GEOINT, which he called an amalgamation of MASINT, IMINT, SIGINT, HUMINT, and more. In 2003, as director of an agency called the National Imagery and Mapping Agency, Clapper set about to integrate all these INT-based sources and tradecrafts into one powerful organization, to be called the National Geospatial-Intelligence Agency, or NGA. This was a defining moment in history. Not only is NGA one of the most formidable and omnipotent U.S. federal agencies at work today, it is also one of the least known. Most people reading this story will have never previously heard of the NGA.

GEOINT is about human activity and where it takes place on Earth. The NGA is a combat-support agency of the Department of Defense. Its partners include all the members of the intelligence community. According to unclassified briefing material, the NGA provides the warfighter on the ground with nine basic geospatial questions to ask, and to know:

1. Where am I?
2. Where are the friendlies?
3. Where are the enemies?
4. When might they move?
5. Where are the noncombatants?
6. Where are the obstacles, natural or man-made; how do I navigate among them?
7. What is the environment?
8. What does it mean?
9. What is the impact?

In war and other conflicts, accurate answers to these questions can mean the difference between survival or death. In the modern era, the U.S. Department of Defense has sought to answer these questions with superior technology systems. GEOINT, or geospatial intelligence, is the latest INT in a series of ascendant steps.

To understand GEOINT in basic terms, consider maps of the American Revolutionary War. When General George Washington needed information on roads, rivers, and fields for planning troop movements, he called upon mapmakers and surveyors to help him and his officers understand local geography and terrain. Decades later, when President Thomas Jefferson spearheaded the government's efforts to possess land held by Native American tribes, he hired explorers like Lewis and Clark to create maps and gather information on tribal leaders out west, an early synthesis of GEOINT and HUMINT.

In 1783, after two French brothers, Jacques and Joseph Montgolfier, succeeded in getting the first hot-air balloon to fly, a new dimension of intelligence collection opened up, from a wide-area, overhead point of view. During the American Civil War, this overhead perspective became a critical military intelligence tool. The Union Army hired meteorologist Thaddeus Lowe to build a balloon that could spy on enemy positions, previously unknowable from the ground. When intelligence from a human operator in the balloon revealed where and how Confederate forces were operating in northern Virginia, the value of overhead intelligence crystallized. During World War I, with the invention of airplanes equipped with film cameras, a new age of intelligence collection was born. America's military pilots could now see and record what was occurring on the ground, from an even wider perspective. Photography sets became highly prized IMINT, with intelligence analysts now able to track how people and objects moved over terrain, across time.

The science and technology born of World War II transformed geospatial-intelligence collection in ways previously unimaginable, paving the way into the world in which we now live. Starting in the 1950s, milestones in science and technology occurred one after the next, beginning with the U-2 spy plane's ability to photograph millions of square miles of Soviet territory from an altitude of 70,000 feet. This terrain had not been seen by intelligence analysts before. Now it could be cataloged, analyzed, and turned into intelligence products. The CIA's CORONA satellite program came next, circling the earth 100 miles overhead, taking reconnaissance photographs of every enemy military installation around the world, and with targetable precision. The National Reconnaissance Office's school bus–sized KH-9 HEXAGON satellite followed in 1971, carrying mapping cameras so precise, analysts who combed over visuals of 877 million square miles of terrain could count Soviet bombers and submarines as imaged from space.

And now, after 9/11, a new age of intelligence collection had been born, one that focused not only on military installations and armies but on individual people. To assign personalized risk. With the invention and fielding of unmanned aerial vehicles like the Predator and the MQ-9 Reaper drones, the ability to locate and track people accelerated at an astounding rate. But with all this imagery came that deluge of data. It was far more information than any one organization could handle. In late 2003, the NGA shared a small portion of its GEOINT database on Iraq with army commanders in the war theater. The army began downloading the data to two hundred of its computers, so the story goes, and army officers became excited to see it. But the warrant officer in charge dampened everyone's spirits by revealing the bald truth: army officers were going to have to be very patient. The download process alone would take an entire year.

Next came a groundbreaking new technology called full-motion video, or FMV, which may sound generic but is not. FMV involves an overhead surveillance system that takes images of a person or vehicle moving through space and time while simultaneously sending geolocation data to a ground control system. This technology was originally designed for precision targeting in Iraq and Afghanistan. As data processing became faster and storage became cheaper, a new methodology called persistent surveillance emerged—the ability to watch and record wide areas where almost nothing happens most of the time, but when something significant does happen, it's being captured on video and digitally stored.

In 2008, a breakthrough in persistent surveillance systems occurred from an unlikely source. The Nevada Gaming Commission had been using wide-area surveillance systems in Las Vegas casinos to catch card cheats. After sharing its "novel surveillance techniques" with geospatial intelligence officers, the Office of the Under Secretary of Defense for Intelligence produced a classified white paper called "Surveillance Employment Strategies for Irregular Warfare." In this paper, "GEOINT analysts found that the same techniques to catch card cheats in casinos could be applied to unravel IED networks in Iraq and Afghanistan," explains Patrick Biltgen, one of the world's experts on activity-based intelligence. The Gaming Commission shared its tradecraft, the data-mining techniques it used to link one criminal to the next, that allowed law enforcement to target and take down criminal networks in Las Vegas. "[NGA] originally called this geospatial multi-INT fusion, or GMIF," says Biltgen, but they eventually settled on activity-based intelligence. INTs are defined by an act of Congress; methodologies are not.

Biltgen is an early pioneer of the classified military technology involved. While employed as a senior mission engineer for BAE systems in 2011, he worked with the data from the wide-area surveillance system ARGUS-IS (shorthand for Autonomous Real-Time Ground Ubiquitous Surveillance Imaging System), one that became part of the Defense Department's Gorgon Stare Program. The ARGUS program was glorified by some for its mind-blowing technological achievements and criticized by others as paving the way for a digital panopticon.

Today, Biltgen has "misgivings" about where society might be heading with this technology. "We're rapidly moving toward a world where everyone is watching [you] all the time. London is full of street-level cameras. A lot of houses in your neighborhood have Ring video doorbells. So you have to imagine in the not-too-distant future, everyone is monitored by multiple sensors everywhere they go and everything they do is recorded in a computer somewhere—just like [the film] *Minority Report* predicted in 2002."

PGSS operator Kevin H has worked in technical intelligence for most of his life. How he wound up in this business has its origin in boyhood inspiration, in a youthful fascination with flight, outer space, and state-of-the-art technology. When he was eight, his estranged father showed up and took him to Las Vegas for a helicopter ride. Kevin H had just seen the original *Star Wars* in the movie theater. "It could have been a ten-minute helicopter ride start to finish, I don't remember, and it didn't matter. It was the most exciting thing that ever happened to me. Changed my life." His father disappeared from his life shortly thereafter, and he only ever saw the man once again. But the experience stayed with Kevin H and set his life on a distinct path, he says.

Many pattern of life subject matter experts who have a background in technical intelligence collection, like Kevin H, work on classified intelligence programs. This means they remain mostly under the radar. Kevin H was moved out of the shadows and into the public domain after he witnessed something at Strong Point Payenzai—activity-based intelligence data regarding First Platoon—that wound up on the desk of the president of the United States.

As a kid riding around in a helicopter over Nevada, that was not something Kevin H could have ever imagined or foreseen. Life cannot be predicted. Except big-data software companies like Palantir believe it can—that actions in the future can be foreseen by using data from persistent surveillance systems.

"By understanding [a person's] pattern of life," says Biltgen, "analysts can build models of potential outcomes and anticipate what may happen."

Activity-based intelligence began in the war theater with the presumption *You are what you do*, but it is now being pushed into a new realm, not just overseas in the war theater but domestically in the United States. It now asserts: *Because we know what you did, we think we know what you are going to do next.*

In Afghanistan in 2012, when not covering down on a specific mission, the PGSS team at Combat Outpost Siah Choy would watch persons of interest like the man in the purple hat. Waiting and watching for three key interactions with the earth that would allow for 429 Status to be assigned to the person of interest. As soon as the criteria were met, the PGSS team would notify the S2 over at Siah Choy. The S2 would monitor the situation by watching the full-motion video feed on one of his forty-inch monitors. At the same time, one of the aerostat's flight engineers would begin reviewing the feed from minutes before.

"Rolling it back in time," Kevin H explains, "in order to take snapshots of the three interactions-with-the-earth events."

The 429 package, which allows an insurgent to be killed in an air strike, must meet the legal requirements. The full-motion video gets snapshotted as evidence. While this is going on, and with eyes remaining on the target, the PGSS operator quickly generates a PowerPoint containing all the data, which goes to the S2 intelligence officer. The S2 quickly reviews that, then sends the information to the battle captain.

"He takes that info," Kevin H explains, "and he washes it through Palantir." Although Kevin H carries a Top Secret clearance, as a PGSS operator he was not able to access Palantir's database. "That's an S2 function," he explains. The job of a pattern of life expert is to find out "who is who" and "who is doing what." They do not have the authority to decide who gets to kill whom.

"The military application of Palantir is awesome," Kevin H says. "Absolutely awesome." Palantir is capable of mining and aggregating data on individual people in a manner that would astonish almost anyone not familiar with its technology. But he thinks the growing movement among law enforcement agencies in the United States to use Palantir's software programs domestically is cause for alarm.

"The fact that there's other moves afoot to actually use Palantir in the United States, I think that's very, very bad, because of the type of 360 [degree] metrics that are collected," Kevin H warns. "I'm not kind of saying, 'Hey, I'm scared of Big Brother.' That's not my view. But that is exactly what Palantir is capable of."

In 2012, there was a geospatial intelligence gap as far as biometrics was concerned. PGSS operators were able to locate and track individual persons of interest who were anonymous. Fighters whose identities were not yet known. Individuals who were being watched because of what they did. Separately, the Defense Department maintained its ABIS database, which contained the biometric profiles on millions of individuals in Afghanistan, some of whom had already been classified as most wanted terrorists. In 2012, there was no technology-based way to bridge this gap. Meaning the MX-15 camera could not positively identify an anonymous person on the ground using biometrics like iris scans. On occasion Kevin H would participate in a go-around.

"I would get a slice of data from Palantir [via S2] saying, 'Hey, this is this guy we're interested in.' The request would be 'Try and locate him.'" Included in the slice of data from Palantir would be an image of the man's face. "I'd get a picture of him," Kevin H says, "old-school, on paper, like a mug shot. I'd also get, maybe, one or two degrees of people that he knows, and areas that he's known to travel in." When Kevin H says "degrees of people," he means individuals the person of interest is linked to, associates with, or has been determined to know. The power of Palantir lies in the connections it can make between people, searching vast amounts of data, analyzing patterns, and making connections that would take humans a vast amount of time to figure out. "Sometimes I'd also get maybe two other guys that are related to the [person of interest] and pictures of them too."

Because Palantir's algorithms could gather data about a person's activities in the past, in 2012, the machines were "learning" how to make predictions about this same person's activities in the future. In addition to the images of the associates, Kevin H would often get predictions about "a general area where [the person] *could* be traveling."

Once the PGSS team located who they thought was the actual person of interest, "we'd kind of do a self-check, to follow him." Meaning the initial hunt began with a computer, but was now fact-checked by a human. "This is basically what I do. I follow his bed-down location. I track every building that he walks to. I determine his daily pattern of life. When does he pray? When does he eat? When does he go to the bathroom? When does he wake up? When does he sleep? The data cuts from Palantir are like a bread-crumb trail for me to go down. At the same time, if I see something, then that's me generating a report. And that becomes data in Palantir." More bread crumbs in the trail.

Once an individual is determined to be a known IED emplacer, like the man in the purple hat, and he has been designated a "429 package," then one of two things happens. "If there is an asset available, if CAS [close air support] is in the vicinity, then it is time to take the target out." If there's not air support available, then the person of interest remains marked for death in the system. "The moment there is a target of opportunity to take him out, I call it in. I don't have to go back through the approving process," Kevin H says. "The 429 package stands. That's why it's called a Target of Opportunity. When you have the opportunity, you strike the target." You kill the man.

One morning, Kevin H came into the ops center. The overnight team was excited.

One of them said, "We're about to kill the man in the purple hat."

Kevin H had personally watched this man bury IEDs. "And I'd watched him train others how to emplace bombs," he said.

Kevin H leaned in close to the screen. "Where is he?" he asked his colleague.

The colleague pointed to the screen.

"Here," he said, "talking to this other farmer," and he pointed to a man seated on a tractor.

Kevin H examined the image feed. The man on the tractor was talking to an old man, who appeared to be another farmer. Kevin H stared at the man in the purple hat. He was approximately 800 meters away from the camera, but the resolution was clear.

"That's a Massey Ferguson tractor he's sitting on," Kevin H said, pointing at the screen.

"Yep," the colleague agreed.

Kevin H explained, in 2020, what went through his mind in 2012. "I'd burned a lot of time and effort trying to locate and kill this guy because he was a [terrorist] cell leader. I knew his face. I knew his gait. I knew his build. I knew what he looked like and I knew he wore a purple hat. I knew he wore white-and-black man-jams [traditional outfit]. I knew the color of his shawl, his little body wrap, and I knew where he lived."

Standing in the C2 shelter at Siah Choy, in front of the video screens, the colleague spoke, "We're getting ready to hit him now," he said. "CAS [close air support] is on the way."

Kevin H stared at the man on the tractor.

"That isn't him," he said. "That is absolutely *not* him."

Kevin H was certain of this. "I thought, Wow, that looks like him. But something just gave me a tickle that that wasn't him. For a lot of different reasons. Number one, he's not a worker. He's a bad guy. Bad guys don't tool around on tractors and play farmer. They are bad guys." The tractor was a legitimate and expensive tractor, one only a farmer would have. "Why is he on a tractor?" Kevin H asked himself. "Why is he talking to this old man in this field?"

The more Kevin H looked at the man in the purple hat, the more he realized something was wrong. "The more I became confused. I said to myself, 'Well, I mean, fuck, it looks like him, but I don't think it is [him].'"

Then he became very stressed out, he recalls. "Hands-down. I wanted the man in the purple hat dead. I still do to this day. But we're talking about killing someone. [Metaphorically], I've got my finger on the button. If that kills an innocent civilian? I don't want that."

Kevin ran out of the operations center, across the outpost at Siah Choy, and into the tactical operations center. "I told the S2 they had to call off the air strike. It's not him," Kevin H told the battle captain. Lt. Grant Elliot, the assistant S2 (intelligence) for the squadron, remembers the man in the purple hat too.

The S2 intelligence officer confirmed that the tactical operations center at brigade headquarters, located a few miles north at Forward Operating Base Pasab, had already authorized the air strike. That close air support was on the way.

"I said, 'I'm certain it's not him.' He [said], 'Well, you've got five minutes to figure that out and prove me wrong.' I said, 'OK, I'll do that.'"

Kevin H ran back to the C2 shelter. "I slew the camera over to his actual bed-down location. He lived right across the river. I waited and waited. It felt like half an hour. It was probably more like a few minutes. Finally he came out. I recognized him right away."

Kevin H was looking at the man with the purple hat. The insurgent who was the cell leader, whose pattern of life he'd been tracking for hundreds of hours.

"He walked out [of where he slept] to go to the bathroom, wash his hands, stretch. I had visual positive identification on him."

S2 called off the air strike.

Kevin H and the team watched as the man on the tractor finished talking with the other farmer and then continued on, farming his land.

"Had a computer done the algorithm on the guy on the tractor, as far as the computer [was concerned], that was him. The [insurgent] in the purple hat," Kevin H says. "But because I had already been watching this guy for months, I knew that it wasn't." Humans are still the ultimate recognizers. "[We] humans have the ability to recognize faces. It's part of our genetics. Of however many thousands of years of being a hunter-gatherer. Of being able to spot recognizable features. I knew his face. I doubted the computer. I was right."

How was the farmer on the tractor misrecognized as the cell leader in the purple hat in the first place? After the air strike was called off, and the man was spared execution, the PGSS operators rolled back the videotape to review what had happened. To see what they could learn.

"It was his hat," Kevin H explains. "There's a [window] of time, around dawn, as the sun comes up," he explains, when colors are "read differently" by the imaging system than how it "sees" them during the day. In this window of time, the farmer's hat was misidentified as purple, setting off a series of linkages that were based on information that was erroneous to begin with.

But what if the S2 shop had killed the farmer in the purple hat in error? And what if, out of fear of backlash over yet another civilian casualty, the data that showed otherwise was deleted so that it would never become known? This invites the question: Who has control over Palantir's Save or Delete buttons?

"Not me," says Kevin H. "That's an S2 function."

Who controls what data gets saved as potential evidence, and what data gets deleted—including data that could potentially be used in a defense? What happens to the rule of law when individual citizens are persistently surveilled without knowledge of, or access to, the information that is being collected on them?

The Department of Defense won't answer these questions on the grounds that its warfighting systems are classified. But as we will see, persistent surveillance systems similar to the PGSS are being used to watch and collect data on Americans back home. These big-data systems are controlled by a select few and understood by even fewer. This exacerbates the potential for inequitable and unlawful actions by the government against citizens because most people do not understand the technology being developed—or they simply accept what they are told the system says is true.

CHAPTER ELEVEN

ABDUL AHAD

By the time First Platoon set up its Alaska tents inside Strong Point Payenzai's fortified walls, the war in Afghanistan had become about the quest for identity. The Joint Chiefs of Staff were now fixated on the idea that "the discovery of true identities" of insurgents in Afghanistan could turn the tide and win the war. To achieve this goal, an all-encompassing intelligence construct emerged, called Identity Intelligence, or I2. "The I2 operations process results in [the] discovery of true identities," the chairman of the Joint Chiefs of Staff declared, as if I2 were some kind of magic bullet. A foolproof way to differentiate between friend and foe, villager and farmer, civilian and insurgent.

For the soldiers of First Platoon, the war was also about discovering identity, but with a meaning that was far different from what the Joint Chiefs had in mind. Identity to the young soldiers was about coming of age. About learning who you are through the brotherhood of the platoon.

"You figure out what you are made of," remembered Private Walley.

"Who your friends are," said Private Thomas.

"Who you are and where you belong in this world," remembered Private Twist.

One of the stranger realities in these two disparate quests for identity involved time. First Platoon would be stationed in Payenzai for less than three months. Many of the insurgents the Defense Department had been tracking around Payenzai had been in the ABIS database going back to 2005.

The discipline of I2 uses data from multiple sources to discover the true identities of people by "connecting individuals to other persons, places, events, or materials [and by] analyzing patterns of life," wrote the Joint Chiefs. Biometrics, forensics, data from documents, even anecdotes acquired from other villagers were all aggregated into intelligence products, one of which is called a biometric intelligence analysis report, a BIAR. These reports are classified Secret or Top Secret and remain inaccessible to public scrutiny, except on rare occasions.

One of the most wanted individuals operating in and around Payenzai in the summer of 2012 was a man who went by the name Abdul Ahad. His BIAR was declassified, and then scrutinized by the author, after the provost marshal general of the army determined its release was a matter of public importance. The discovery of this man's true identity—not made public until this book—reveals as much about the Defense Department's classified biometrics programs as it does about the future of the rule of law in the United States. Abdul Ahad plays an extraordinary and very strange role in the tangled narrative of First Platoon.

Abdul Ahad first came to the attention of the Defense Department on March 11, 2009, somewhere in Zhari District, Kandahar Province. Two different versions of the man's original capture and detainment are recorded in his case file, or BIAR. In one version, a counter-IED unit was moving toward an observation post when it received radio traffic indicating troops were taking fire. An insurgent was shot in the exchange, and the record states: "1 x enemy wounded in action." The counter-IED unit located and recovered two antitank mines with detonation cord attached. Also recovered was an "ICOM scanner"—a handheld radio used by Taliban insurgents to communicate with one another—a cell phone, and $3,900 in Pakistani currency. The large sum of foreign cash suggested to intelligence analysts that the fighter held a leadership position in a terrorist organization. In Afghanistan, the average annual income in 2009 was around $200. After being detained, the man's name was recorded as Abdulhat Abdulhat.

One of the members of the counter-IED unit took a buccal swab of the terror suspect's cheek, collecting cells for a DNA sample. The bomb-making material was collected, bagged, and tagged as evidence. It was sent to the army's forensics laboratory at Bagram Airfield for analysis. In 2009, this laboratory was still officially called the combined explosives exploitation cell, same as the ones FBI Special Agent Scott Jessee helped the DoD set up in Iraq.

The man's DNA sample was sent back to the United States, to the Armed Forces DNA Identification Laboratory. Based on this biological specimen, the man was assigned Known DNA Profile 2009X06332-0001A, an alphanumeric sequence that would be used to identify and track him for as long as he was alive.

A second declassified version of events describes an alternate version of how the man was detained, this one less kinetic. "The subject walked into a CF [Coalition Forces] observation post carrying 2x antitank mines with detonation cord, 1 x ICOM scanner and 1x cell phone; CF detained the individual at the [outpost] until the counter-IED element arrived."

During the war in Afghanistan, it was not uncommon for a villager to walk up to an outpost and say he wanted to turn in an IED in exchange for a financial reward. The army encouraged it, paying between $50 and $100 per IED. This program was part of former ISAF Commander Gen. David Petraeus's original "money as a weapons system" strategy, the underlying principle of which was that the army could transform Taliban sympathizers into pro-U.S. actors with force of money. But in Afghanistan, the IED turn-in program also led to problems. It became a means by which Taliban commanders conducted reconnaissance on American bases. Posing as a villager afforded an insurgent quick and easy visual access to base security measures, paving the way for a later ambush or attack.

Which version of the man's capture was accurate remains unknown. The end result was the same: he was detained and taken to a larger forward operating base in Sarkari Karez called FOB Ramrod. Here, the man's full biometric profile was recorded using the BAT and HIIDE systems. The record indicates that a soldier took scans of the man's ten fingerprints, two palm prints, two iris scans, and three facial images. Additional recorded information indicated the man was five-foot-three, weighed 113 pounds, and had black hair and hazel eyes. He was from the Pandaahkah tribe and he had been born sometime in 1974, which would have made him thirty-five years old in 2009. In the declassified facial images taken of him, the lines on the man's weathered face make him appear much older, possibly fifty, or even sixty, years old.

Two days later, on March 13, 2009, Abdulhat Abdulhat was transferred to the Afghan National Directorate of Security. In existence since 2002, the NDS is Afghanistan's domestic and foreign intelligence agency, analogous to the FBI and the CIA in one entity. Notoriously corrupt, in 2009 the NDS had a reputation for acting with impunity. For torturing some detainees, letting others go free, and using still others as paid informants. According to his BIAR, Abdulhat Abdulhat remained in NDS custody for a period of four days. There, a new piece of information emerged: the man's name was Abdul Ahad and he was the son of a man called Haji Payand from Zhari District.

On March 17, 2009, Abdul Ahad was turned back over to U.S. forces. From there, he was taken north, to the Detention Facility in Parwan, the locus of the Koran burning fiasco. If you recall, this is the American-run prison located just west of Bagram Airfield. In the rule-of-law paradigm, the interplay between law enforcement, courts, and corrections was foundational. The guiding principle of the rule of law—that all people and institutions are subject to, and accountable to, the same set of laws fairly applied and enforced—depends on these three systems working together. In Afghanistan, because the country was a war zone and Taliban insurgents were not going to post bail, a fourth element had been added: detainment. And so, Abdul Ahad was taken to the Detention Facility in Parwan, where he remained, detained.

Nine months later, on December 31, 2009, Abdul Ahad's biometrics were again recorded. The digital data remained the same, but the narrative information changed again. This new record indicated Abdul Ahad was one inch shorter, five-foot-two, weighed twenty-six additional pounds, and had black eyes instead of hazel ones. The alleged height loss could be accounted for by quickly taken measurements; the weight gain by hearty prison food; the different color eyes perhaps because of lighting. Abdul Ahad gave no other names, still said he was from the Pandaahkah tribe, but now stated he was fourteen years older, that he was born in 1960, which would have made him forty-nine or fifty years old at the time. Both of his parents were dead, he said. The soldier recording Abdul Ahad's biometrics noted that he had tattoos on his wrist. One key difference: this record stated that Abdul Ahad was blind in his right eye. His fingerprints, iris scans, palm prints, and DNA were all the same, forever in the Defense Department's ABIS database, which was growing rapidly in power and in size. Abdul Ahad was assigned the alphanumeric biometric identification (BID) B28JM-UUYZ by the Defense Department, a number that he could be positively identified by for future match-hits.

Meanwhile, the IED components recovered with Abdul Ahad in March 2009 were sent to a classified forensics laboratory at Bagram Airfield. There, behind a maroon door accessible by a key-pad code combination, a team of twelve soldiers and three contractors worked around the clock as part of the combined explosives exploitation cell, Afghanistan, or CEXC-A. This team was made up of experts trained in combat biometrics, intelligence, electronics, explosive ordnance disposal, and photography.

When the IEDs associated with Abdul Ahad first arrived here, EOD technicians examined the bomb components to make sure there was no remaining explosive material. This included sending the material through an X-ray machine. After the IED was rendered safe, it was photographed from various angles, as were all the ancillary items, including the wires, the detonation cord, the tape, and the batteries. Every item was documented, inventoried, and photographed, then entered into a database, where it was assigned a bar code so it could be tracked.

There were three labs within the CEXC-A at Bagram: one for biometrics, one for electronics, and one for intelligence. Lab technicians in the biometrics lab pulled latent fingerprints off the IED parts of the bombs associated with Abdul Ahad, using different procedures for different materials. When forensic experts searched for latent fingerprints, they handled porous items like cardboard, paper, and wood differently from how they handled nonporous items like batteries, wires, and electrical tape. Meanwhile, technicians in the electronics lab evaluated how the IED might have been constructed. They searched for what are called signature techniques, clues that might affiliate this individual with already-known IED manufacturers in Zhari District. Analysts in the intelligence lab scoured the reports, looking for leads. Each person on the team shared a common goal: determine who was the bomb maker so as to remove him from the battlefield.

The CEXC-A team at Bagram worked on gathering forensics and collecting evidence that could be used in a criminal case against Abdul Ahad. Halfway around the world, in the United States, an assemblage of other efforts was under way. At numerous federal facilities, teams of biometrics experts were now running Abdul Ahad's fingerprints and DNA against federal databases to try to locate a match-hit. In 2009, these efforts were managed by the Biometrics Identification Management Agency, the new name for the Biometrics Management Office, originally run by John Woodward in the early days of the global war on terrorism.

Eleven months passed. On November 17, 2010, a unit of coalition forces were on patrol near Chaharshakha Village in Zhari District when they were "engaged" by a radio-controlled IED. The declassified record does not indicate if any of the soldiers were maimed or killed, only that a second pressure-plate IED was also discovered, and then recovered, by an EOD technician. Three 9-volt batteries, one roll of tape, one lightbulb, and a quantity of wire associated with the homemade bomb were documented, bagged, and sent to a new forensics laboratory, one that had been set up locally at Kandahar Airfield.

Colonel Ed Toy, the man who built this first forensics laboratory in Kandahar, described to me in 2019 what this effort entailed. Colonel Toy served as chief of counter-IED operations in Regional Command South and deputy commander of Joint Task Force Paladin at Kandahar Airfield. Building, and then running, a forensic-science laboratory in a war zone, he says, was not something he will ever forget.

"You have to remember the whole place was like a scene from one of the *Road Warrior* movies with Mel Gibson," Colonel Toy said.

"So imagine doing [forensic science] in that world. At that time, we didn't have a lot to work with. I'd walk by rows of forty-foot containers bursting with evidence . . . clear plastic bags, pressure plates, and all kinds of bomb components brought in by units that had come into contact with all these IEDs."

Bomb makers leave behind clues. "It may be how the bomb maker twists a wire or solders components to a circuit board . . . if the bomb maker is a seasoned builder who has made hundreds of bombs, they will develop tendencies unique to them," he explained, referring to the bomb maker's signature. "Other bomb makers are careless and sloppy in their work, leaving behind easily identified fingerprints and even hairs from their beards." This work is labor-precise and labor-intensive, he says, and requires documenting the evidence, writing reports, packaging and repackaging the evidence. "We had to repeat this process dozens of times a day, every day . . . in a virtual factory of forensic evidence that would make any automotive-plant assemblage seem tame."

Colonel Toy's work took place between 2008 and 2009. As hundreds of millions of dollars were pumped into defense, new programs were being introduced at a rapid rate. Starting in 2010, there was a new emphasis on DNA. In the twenty months since Abdul Ahad was first detained with two antitank mines and $3,900 in Pakistani currency outside an outpost in Zhari District, the science behind battlefield DNA testing and matching had advanced exponentially, owing to the DoD-funded efforts of a microbiologist named Dr. Richard Selden and his team. In November 2010, the IED components captured near Chaharshakha Village were swabbed for immediate processing, to be looked at locally in Kandahar, in a fancy new mobile DNA lab set up there.

The army's mobile DNA labs in Afghanistan were a sight to behold. Portable and deployable, each module was roughly the size of a shipping container and outfitted with tools as commonplace as superglue and as rare as a $50,000 laser. "These [labs] have the same capability as any major crime lab in the U.S.," the army's top DNA examiner, unnamed, said for an army press release. Many of the forensic scientists forward-deployed to Afghanistan came from the Defense Forensics Science Center in Atlanta, Georgia. With every precaution in place to prevent cross-contamination, the DNA examiners in Kandahar performed their work to the same high standards as in the United States, the army said. They wore lab coats, face masks, and gloves, and kept work surfaces cleaned with bleach. "It's a big deal because once a sample is contaminated, it's done."

For DNA examiners combing through bomb components found in Chaharshakha Village, the hunt was on. They searched tape and wire for any usable touch DNA, just a few human skin cells or a single hair would do. Once a specimen is obtained, the samples go through several different lab processes, "with the end result showing a graph that marks the DNA sequence at 16 locations along the molecule."

For decades, this multistep process had taken months. Examiners had to process the sample, evaluate the raw data, write a report, and submit that report for peer review. It was the last step that was often the most time consuming, because the peer-review process is historically rigorous. One "where at least two other forensic scientists [must] agree with the results," the army says. But on the battlefield, there was little time to wait. Combat biometrics had sped up the process through automation. In 2010, in the mobile labs, DNA results were being completed in as little as thirty-two hours, far faster than the five days it took to test Saddam Hussein's DNA. Once the results were in, the DNA profile could be compared to millions of other DNA profiles the Defense Department kept stored in its database.

Six weeks later, on January 18, 2011, forensic analysts with the Armed Forces DNA Intelligence Laboratory back in the United States got a match-hit. DNA obtained from the IED discovered on November 17, 2010, matched Known DNA Profile 2009X06332-0001A.

A mind-boggling amount of data, aggregated by machines and then analyzed by humans, had led to the confirmation that this IED had been built by Abdul Ahad. This was precisely how the multibillion-dollar system—the Manhattan Project–style biometrics program proposed in 2004—had been designed to work. The evidence was incontrovertible: "DNA match was obtained from a wire twist clipped from an IED and matched to the subject's known DNA sample collected while he was detained." Abdul Ahad was a terrorist bomb maker. Unmasked by science, he was no longer anonymous. *Fingerprints don't lie*, so the saying goes. "DNA is 99.9999999999 percent accurate," says Dr. Richard Selden.

Abdul Ahad had been detained on the battlefield in accordance with the rules of engagement. Forensic science linked him to a crime. Afghan lawyers at the Justice Center in Parwan—located next door to the Detention Facility in Parwan—could now use forensic evidence against him in a newly set-up Western-style court of law. If convicted of terrorism crimes, Abdul Ahad would be sentenced by an Afghan judge, to be incarcerated in an Afghan prison. Except for the fact that eighteen days prior to the DNA match-hit, Abdul Ahad was released from Afghanistan's Guantánamo Bay on New Year's Eve. No statement in his biometric intelligence analysis report as to why. All the U.S. Army could do now was wait to cross paths with Abdul Ahad again.

Two months later, another find. Afghan police were "targeted" by an IED, the record shows. Again, no indication if anyone was killed or maimed. Following protocol, the evidence was collected, bagged, tagged, and again sent to the lab in Kandahar. This time, a second DNA match-hit linked the bomb to Abdul Ahad. Abdul Ahad was not just a Taliban bomb maker, he was the leader of an IED cell. "The subject is assessed to be an IED manufacturer, and may possess knowledge regarding emerging tactics, techniques and procedures, to include [IED] construction methods." His level of personalized risk was elevated. He was a most wanted individual now.

At the laboratory in Kandahar, biometrics specialists worked an IED Identification Link Analysis diagram on Abdul Ahad. This "spider chart" determined whom he reported to; who reported to him; how often he traveled to Pakistan; the weapons he bought and sold; and the Coalition Forces bases he'd attacked in the past. Analysts determined his bed-down location and, as revealed in declassified documents, marked this identifying location for him with crosshairs on a map. By the winter of 2012, Abdul Ahad was placed onto the Joint Prioritized Effects List, or JPEL, a highly classified list of persons the Defense Department intended to kill in an air strike, or capture by U.S. ground forces, whichever happened first.

In the global war on terrorism there were, and remain, numerous kill lists, each guided by a different set of classified and unclassified legal constraints. The JPEL, which was run by the Joint Special Operations Command, or JSOC, was seen as particularly controversial because of its omnipotent reach. John Nagl, a former counterinsurgency adviser to General Petraeus, described the JPEL to *PBS Frontline* as "an almost industrial-scale counterterrorism killing machine."

If First Platoon were to stop Abdul Ahad or any of the insurgents linked to him through Identity Intelligence, these most wanted terrorists would be detained. That was the idea.

CHAPTER TWELVE

THE THREE IEDS

Private Walley smoked menthols, and he always had a robust supply. Earlier in the deployment, at Kandahar Airfield, he bought ten cartons after he learned they were only $5 each. He never ran out the whole time he was in Afghanistan because the only other person who smoked menthols was Specialist Brian Bynes. "Smoking was almost mandatory," Walley remembered. "If you didn't smoke before, you probably took it up there." Specialist Alan Gladney didn't smoke cigarettes, but he smoked a pipe.

On May 30, 2012, First Platoon was assigned to drive Second Platoon to a mission down by Siah Choy. Privates Walley and Twist were two of the soldiers who drove them there, in armored vehicles. It was an approximately 2.5-mile distance from Ghariban to Siah Choy as the crow flies, but more like four miles on the roads engineered by the U.S. Army. The main supply road from Highway One down to Siah Choy was loaded with IEDs. This route was named Route Victoria, but the PGSS team called it Route Vicki.

One of their jobs was to keep brigade commanders updated as to where IEDs were buried along this road.

"All the IEDs were all marked as data points on the map," Kevin H remembers, referring to the geospatial intelligence product they were regularly updating. Wherever anyone on the PGSS team saw an insurgent bury an IED, he says, "we'd mark it up and send the [data] over to the S2, where it becomes [part of] one big, gigantic plotted map."

The fact that the army had precise coordinate information indicating where specific IEDs were buried in every platoon's area of operations, including IEDs that would maim and kill its own soldiers, has largely avoided public scrutiny.

"The intention was to get the information about the IEDs to the Joes," Kevin H explains. Before stepping off point on a foot patrol, an NCO or a lieutenant "would say, 'Hey, we're going down this path. We're going to visit this village,' and we'd send word back: 'Watch out for these marked IEDs.' But for whatever reason," he says, "some information gets ignored. Or isn't retained."

None of the NCOs interviewed for this book recall the reporting system working in this manner. "I only [vaguely] knew about airborne assets like the PGSS because of a previous deployment where I worked in a tactical operations center," says Sergeant Williams.

Kevin H tells a nightmare story.

"I'd seen an eighty-pound IED that I knew about for eight months." Day after day, it was avoided by the platoon assigned to patrol the area. Then, "in came a new platoon. I come into the C2 and [through the surveillance camera] I see them headed right toward it. I'm like, 'Hey, stop these guys! Get on the horn with S2. You've got to stop those guys, they're going to walk over an—!' Too fucking late."

Kevin H does not provide more details about what it is like to watch an American soldier step on an eighty-pound IED.

"Unfortunately, you have to get used to it," he says, "to do the job. Your job [is] to keep trying to get the information to the battlefield. So when a platoon steps off, you say, 'Hey, if you go five meters west down this path, there's going to be an estimated twenty-pound IED there.'"

But there was simply too much data to manage. And too many competing agendas. The army was "swimming in sensors and drowning in data," just like the general said.

When certain things happened, PGSS operators knew to mark notes in their calendars. Certain things "lawyers [might] want to know," Kevin H clarifies, "like possible CIVCAS," civilian casualties. A U.S. soldier getting hit by an IED was something "to expect." If it sounded callous, it was war.

Captain Swanson remembers what seemed like double standards coming down the chain of command, from brigade commander Col. Brian Mennes at FOB Pasab. "Colonel Mennes's attitude was, if we got hit by an IED, we'd been 'outmaneuvered by the enemy.' Those were his words."

Sergeant Herrmann said, "You tell the privates, 'Pay the fuck attention.' You drill down on them before you deploy, 'Not everybody comes home.'"

Walley and Twist drove the soldiers from Second Platoon to the mission launch site, down near Siah Choy. Private Walley recalled having a bad feeling about the area. "It was really early in the morning. Still pitch-black, before dawn. I could see women and children peeking out from behind doorways. It felt like a scene from a horror movie, there was even some kind of fog."

For Twist, "The whole area felt cursed."

The closer you got to the river, the more people. The more people, the more invisible threats. Second Lt. Jared Meyer served as platoon leader for Second Platoon. He recalled the overarching mission that Colonel Mennes laid out for the platoons: "The intent that we set out to achieve was 'stir the pot.' Shake it up. If necessary, take fire. With the ROE being what it was, we couldn't just go out and detain people. We had to get them to shoot at us before we could take any action or we had to catch them doing something or we had to have an informant. All of those things are achieved by being out [on patrol] and stirring the pot."

From an overhead surveillance perspective, and from higher up the chain of command, the mission had other goals as well. Army intelligence wanted information on an abandoned mosque in the area, one that the Taliban was using for torturing people. "We'd spent hours covering down on this place," Kevin H says. "It wasn't really abandoned anymore because the Taliban had taken it over. They'd grab locals and train them to plant IEDs. If they resisted, they'd hook them up to a car battery. That kind of thing."

S2 wanted images of the area, from the ground. Digital photographs of places and spaces that could not be seen by the cameras on the PGSS system from overhead. Places inside compounds, around corners, behind walls. Data on persons of interest in the village: fingerprints, iris scans, facial images, and DNA all gathered by the COIST, to be uploaded into the ABIS database for future use.

"ROEs said we can't engage," Lieutenant Meyer explains, "so going out on patrol [was] the roundabout way of achieving Colonel Mennes's goal." Invariably the platoon would draw fire, which allowed them to return fire legally, according to the rules of engagement.

The area had proved lethal for a platoon sent into it the month before. Kevin H recalls the mission because the PGSS team watched it on a live feed and a sergeant named Nicholas Dickhut was killed. Reuters photographer Baz Ratner was embedded with the platoon and took striking photographs that were printed in newspapers around the world. One of the images drew an abundance of comments with readers expressing shock at how anachronistic the war in Afghanistan seemed. There were American soldiers, outfitted in the most modern accoutrements of war, surrounding an indigenous barn used by local farmers to turns grapes into raisins, a structure called a grape hut. With its hand-hewn mud-brick walls, irregular-sized windows, and crooked tree branches poking out of the sides, the grape hut looked like it belonged in the Stone Age. The PGSS team knew the Taliban used the structure to fire at soldiers, and to hide weapons caches. Baz Ratner took a photo of Sergeant Dickhut pulling security inside the grape hut after the soldiers took control. Not long thereafter, Dickhut was dead, killed by Taliban trying to take back their barn.

"We named this particular grape hut the Grape Hut of Woe," Kevin H recalls.

Now Second Platoon was heading toward it again, to recon the area and gather biometrics on the villagers. This time, the army took a different tactic on approach. The patrol walked behind four 70-ton Caterpillar D9 armored bulldozers, engineered for battle and with slat-armor caging to deflect rounds from rocket-propelled grenades.

"They were bulldozing what seemed like a five-lane highway," Lieutenant Meyer recalled.

The terrain leading up to the village reminded Meyer of a Midwestern town, like in Iowa, near where his dad lived. "Great big fields." Land wide and flat. The soldiers patrolled forward, taking ground slow and steady, following the bulldozers one step at a time, sweat-soaked. Determined. Unnerved.

At the edge of the village, the bulldozers reached an impasse and were unable to proceed. Lieutenant Meyer ordered the soldiers to continue moving forward, on foot. Specialist Nicholas Olivas was out in front with him.

Meyer recalled, "He'd taken maybe twenty-five steps beyond where the bulldozer stopped. He looked at me, as if he was asking, *Should I go left or right?* I nodded right. He took one step. That one step—"

Every soldier's nightmare.

"Nick stepped on an PPIED, buried right there in the dirt."

The explosion created a blast wave that traveled outward at a rate of approximately 1,600 feet per second. One thousand miles per hour. The blast knocked everyone in a twenty-five-foot radius back, up, out, and down. Sent projectiles flying. Left lungs, ears, abdomens, and organs wracked with varying degrees of overpressure damage. Rattled brains.

Nick Olivas lay flat in the dirt, his left leg sheared from his body just below the hip, his right side shredded and torn. There was blood everywhere. Platoon mates rushed to him and began tying tourniquets on the stump. After a brief blackout, Meyer regained consciousness.

"I took Nick's hand and began telling him everything was going to be OK."

Nicholas Olivas from Fairfield, Ohio. Twenty years old. A high school wrestling star. Married to Faith Compton Olivas. New father of Connor, a five-month-old baby boy. Son of Adolfo Olivas, the town's former mayor, and Marion Olivas, a sheriff's deputy. Now he was dying. One of the soldiers used the nine-line to call a MEDEVAC. The others kept pressure on Olivas's wounds.

"We got this," Lieutenant Meyer recalled saying to himself. The helicopter was on its way. At the rate things were moving, Olivas would arrive at the trauma center in Kandahar well within the golden hour. This mandate had been set by Secretary of Defense Gates—one hour from injury to operating table—and it was saving lives. But Olivas would die on the operating table. When surgeons unclamped his femoral artery to operate, it got sucked up into his body and Olivas bled to death.

"We all thought Nick survived," remembered Meyer. "We continued the mission." Notification comes from higher up the command, and Colonel Mennes didn't want the soldiers distracted by a platoon mate's death. News came at the end of the following day. By then, Second Platoon had set up a tactical operations center inside the Grape Hut of Woe.

At Combat Outpost Siah Choy, the PGSS operators covered down on the platoon, pulling megabytes of data by the minute and waiting for more data to come in—information that could only be obtained on the ground, by the COIST: digital imagery of buildings, biometric data from villagers, forensic data from weapons caches. At Strong Point Payenzai, Sergeant Williams was given an order to take a group of soldiers down to Siah Choy in an armored vehicle, to pick up what was left of Olivas's things.

The drive was silent, he recalled. The armored vehicle had a road grader out front, designed to absorb the blast if they hit an IED. When Williams spotted what they'd come for, he felt sick.

"It was all in a black garbage bag," Williams remembers. "The privates all knew what it was, but no one said anything. They were all friends with Nick."

Private Walley remembers feeling slayed. "I cried with Reynoso for about thirty straight minutes. Then I went numb."

Olivas's death cast a long shadow over First Platoon. Every patrol, in every village, had a menacing feeling. On June 1, Private Thomas turned nineteen. Only two more years until he could legally drink. He got grabbed by his platoon mates and belly-slapped by Private Shilo nineteen times.

June 4, 2012, was Walley's twentieth birthday, the day the platoon took biometrics on the second most wanted Taliban in southern Afghanistan. Captain Swanson vaguely recalls processing this detainee. In point of fact, capturing the number two Taliban was not as rare as it sounds, he says. "There is always another fighter" who moves up the chain of command to assume the number two position.

June 6 began like the other days. Wake up at 4:30 A.M. Drink some water or a Rip-it. Eat a meal ready to eat, an MRE. At the pre-briefing, the soldiers learned they were going to recon the area around Mullah Shin Gul, an abandoned village 700 meters north of the strongpoint. There was a two-story compound there, shaped like an *L*. A known bed-down location for the Taliban.

"I hated that they kept making us go back there," recalled Sergeant Herrmann. "There was nothing there." Except more forensic evidence from another crime scene.

Approaching Mullah Shin Gul, Twist carried the Thor on his back. He was monitoring the area for radio-controlled IEDs. Close air support checked in and reported seeing insurgents, possibly carrying weapons, hiding in a field. Sergeant 1st Class Keith Ayres found a weapons cache in a culvert and called for the COIST to bag and tag the evidence, for fingerprints and DNA. A man on a motorcycle came driving into the area. Ayres stopped him. The COIST took his fingerprints and iris scans. No match-hit. But the villager was jittery and seemed suspicious.

"Ayres told him to stay the fuck away," Sergeant Williams recalled.

While Ayres dealt with the cache and the COIST, weapons squad walked on toward the L-shaped compound.

"Fire started coming; we returned fire," Williams recalled. Then things got quiet again. Private Walley was ahead, carrying the Squad Automatic Weapon, the SAW. He'd switched positions with Haggard the week before. Haggard was near the rear of the platoon now, working his way up and over the grape rows.

"The compound had wooden bars on the windows," Walley remembered. "Not once did I see anything like that on any window in Afghanistan." Wood was expensive. A prized commodity in a poor village, and rare.

Walley was out in the open now. There was a path big enough for a motorcycle. Uneven terrain.

"We made our way up a hill, down, and around to the bottom

back door of the mud compound. I cleared the small part" that acted like an entrance, Walley recalled. One of the NCOs sent 1st Sergeant Joseph Morrissey, Specialist Reyler Leon, and Private Twist to another part of the *L*.

Walley made his way into a small room inside the compound. He was alone.

"I saw burn marks in the corner, on the wall," he recalled. "Thought, Oh, they're making IEDs. There was a dull knife. The kind used to cut HME [homemade explosives]. I came out of the room. There were wires on the wall. Wires on the door."

Walley called out to Sergeant Williams. Said, "Get over here, now."

As Williams approached, he saw it too. Wires on the wall. Wires on the door. "We were standing in the middle of an IED factory," Williams remembered.

Morrissey and Leon had the minesweepers and got called away.

"It was Twist, me, Sergeant Will," Walley remembered. "The three of us were right there."

Walley heard a motorcycle coming. Then he saw it too.

"I looked at Sergeant Will. He nodded at me. *Go kill him,* his look said."

Anyone who was here in this IED factory was a threat and could be neutralized according to ROE. Walley remembers trying to gain situational awareness, and that there were almost too many visuals to deal with. Wires everywhere. Was that a block of HME?

"I looked back at Sergeant Will," Walley remembered, trying to keep track of the rider on the motorcycle. "He nodded again: *He's in the kill zone.*"

Walley moved toward the man.

"I'm going to kill him," Walley recalled thinking. "Then . . . they detonated. The IED went off right in front of me."

Twist was right there, closest to Walley. He remembered, "It was a bright sunny day. Bright sky. Dry day. I was looking at [Sergeant] Will—"

The blast blew Twist back onto a wall. "There was ringing in my ears. It was dead quiet. What just happened? Sam was screaming. I realized he was bleeding immensely. I thought, Oh, this isn't good."

Twist set down his weapon.

"*Do something.* What just happened? Screaming. *Well, go get to Sam,*" Twist remembered thinking. Then he began running. "I didn't really understand what was going on."

Twist ran toward Walley. Ten, fifteen, twenty feet across an open area.

"*Get him out of the ditch,* I thought. It still didn't register what was going on. I grabbed him by the back of the vest. I saw his left arm was toast. His right leg was just the tibia and fib, that was it. Degloved to the bone. It looked like a knife at the end. I know there's a leg there, I thought. Part of a leg." Twist's brain said hurry up. The smoke and the dust were clearing now. "*Get tourniquets on,*" Twist told himself. He carried a lot of tourniquets. "In fact, I carried more tourniquets than the average bear. And I just remember thinking, *Get tourniquets on your friend.*"

Twist bent down. As he was tugging Walley out of the hole he thought, *Oh, there's his foot. And his boot. All by itself. Way over there.*

Blood was gushing out of Walley's amputated limbs. Twist got on three tourniquets, quick. "I was handling it. But emotionally I took it hard. There's no training for what happens when one of your buddies steps on an IED. Sure, there was training with pigs at Bragg. But they can't really simulate mass hemorrhaging. The weird thing is, I broke down in the moment and I also carried on."

The IED had been deeply buried in the earth and now there was a crater, four or five feet deep. Six feet wide. Twist finished dragging Walley out of the hole. Walley was awake and alert but at first he couldn't see. Then he could.

Walley had not lost consciousness. "I watched Joe [Morrissey] Olympic-hurdle over a grape row and come running." He realized Twist was over him, looking down at him now.

"By looking in Twist's eyes I thought I was dead," Walley remembered. "Then I realized he was alive."

Sergeant Williams had Walley's head in his lap and he was comforting him. Twist kept tying tourniquets above Walley's missing limbs. Everything was covered in blood.

Walley moaned, "Oh God. Why me, why me, why me?"

Sergeant Williams remembered, "He kept on repeating *why me?* I didn't know what to say, so I said, 'I don't know.'"

There were lots of trees and grape rows. Always, Afghanistan's bright blue sky overhead. And the knowledge "there was probably no way a MEDEVAC could land," Walley recalled.

"I prayed," he remembered. "Please, God. Get me out of here. . . ."

Sergeant Williams used the nine-line to call the MEDEVAC. Walley's right leg was mostly amputated, and the bottom part of his left arm was gone too. What remained of the left leg was mangled and ripped apart. Williams called it in as a "triple amputee."

All these years later, Walley says he still thinks about the MEDEVAC pilot who landed in such a small area so flawlessly, and who no doubt helped save his life.

"I watched rotor blades clipping trees. I knew Olivas hadn't made it and I thought, I'm not going to make it. I'm probably going to die."

The chopper landed. Twist kept tying tourniquets and wouldn't let go.

"We loaded Sam in," Williams recalled.

Twist tried getting in the helicopter with his friend Walley. But some of the other soldiers pulled him back and wouldn't let go. The helicopter took off, for Kandahar.

"What just happened?" Twist remembered thinking. "Did that actually happen?" Twist wondered in 2019. Later, "I would ask myself this question over and over and over. Sometimes in the middle of the night. Trying to solve the puzzle of what happened [that day]." PTSD is like that, Twist figured out years later. "I would sit there going over everything I would remember, in my mind. Sometimes in the middle of the day. Sometimes all night. What happened? How did that happen? Why did that happen?" Questions on a loop. "It never goes away."

Walley was in the bird. The soldiers needed to get back to Ghariban.

"We had an enemy prisoner with us," Sergeant Williams remembers. "We walked him back." The COIST took the prisoner's biometrics and uploaded them into the database with the information from the weapons cache, the motorcycle rider, and photographs from the area where Walley got hit.

Later, after Sergeant Williams had some water and a moment to rinse out the blood from his clothes, Captain Swanson showed him photographs of the IED crater that had been taken by the COIST.

"There was a big crater. I saw Sam's boot. Part of his foot was still in it. Part of his leg."

On the flight to Kandahar, Pfc. Samuel Walley kept touching the medic's face. Inside the trauma center, on the operating table, he remembers staring into the eyes of the doctors. Their eyes were large and oversized, "peering down on me behind these surgical masks." He remembers pain. That he had no anesthesia because his heart rate "was already way too low." What happened in the trauma center at Kandahar Airfield, says Walley, "was the most horrifying thing you can have happen to you as a person. They cut off the rest of my arm. Threw it in a biohazard [bin] like a piece of trash."

Walley getting hit by an IED lengthened the dark shadow over First Platoon. Twist remembers thinking what was going on was all for shit. Williams remembers thinking it was up to him to make certain the soldiers soldiered on. Every day, an endless loop. Presence patrols. Honesty trackers. Get out and meet the villagers. Let them know we're here to help. Host key leader engagements. Here's some free medicine but give us your fingerprints and your iris scans first.

Reynoso did more pull-ups. Bynes smoked Walley's menthols. Specialist David Zettel and Matthew Hanes, promoted to specialist by now, strung a hammock and read Hemingway. Private Thomas and other soldiers played dominoes, and also cards, with imaginary money, keeping track of who owed whom what, on a chit. Someone set up a game of horseshoes. It was absurd, yes, but someone else's mother sent the lawn game to them and those weighted, iron horseshoes were a link back home to America. To a life waiting for each of them there. Sergeant Herrmann shut himself off in the air-conditioned MATV and read, and read—sometimes all night long.

June 13 began like every other day. Get up at 4:30 A.M. Line up at the entry control point. Drink some water or a Rip-it. Eat an MRE. Get ready for another presence patrol, this time into Payenzai, a few hundred meters south. Payenzai Village: where 2nd Infantry Division shot the kid.

"We'd seen a guy digging in the service rows that connected the grape rows to the road," Specialist Fitz recalled. "Lieutenant Latino [needed us] to go there to see a village elder."

The patrol avoided the road, laden with IEDs as it inevitably was. They crossed a small dirt area, passed a marijuana field, and headed into the grape rows.

"Everything smelled like anise," Fitz remembered. "Anise grows wild and it smells like black licorice. If I smell it today, it makes me sick. Burnt nitrate and anise. That's what everything smelled like around Payenzai. Once I drank anise liquor after I got home. I immediately threw up."

Fitz was in front of Private 1st Class Mark Kerner. Behind Kerner was Lieutenant Latino, and behind Latino was Private Thomas. Every soldier walking ten to fifteen feet apart.

"We were inside a grape row when Kerner stepped on something a bunch of guys had just walked right over," remembered Thomas. "I heard a loud pop."

Specialist Haggard heard the pop too. "Somebody says, 'Fuck.'"

"The fuse [might have been] wet, which made a delay," Fitz recalled.

Kerner just stood there, his foot on a PPIED.

"Everybody froze," remembered Thomas.

"I cradled my M14 in place," Fitz recalled. "I was expecting a blast. I looked Kerner right in the eyes as he said, 'What the f—'"

The blast went off. Private Thomas got thrown onto a grape wall and knocked unconscious. Fitz too. "I remember coming to, laying in the grape row," said Fitz. "I had the aid kit and the litter backup. I was a few feet away from Kerner. I was really scared to move."

Private Thomas remained unconscious for maybe a full minute or two before he, too, woke up. "Everyone was moving all around."

Kerner was missing a large part of his backside. Lieutenant Latino was bleeding from the neck and face.

"He had a large piece of metal protruding out of his neck. Maybe a centimeter or two from his jugular," remembered Haggard.

"No one knew exactly what to do," remembered Fitz. "Then [our medic] Doc Rose just walked up and pulled the metal straight out of Latino's neck."

Private Thomas was still on his back, looking up at the sky. There was a really loud ringing in his ears and he couldn't hear what anyone was saying. "First there was silence, then the super loud ringing, then the ringing faded into slurs, into people talking. Into kind of being able to make out words." Thomas stood up. Looked at his arms and his legs. "I realized I was completely OK." Pieces of Kerner's backside were scattered around. Blood on the grapevines. Burnt leaves.

Everyone was doing something. It was Thomas's job to radio Ghariban.

"Troops in contact!" he yelled. Then sent coordinates for their location.

But Ghariban needed better coordinates, they said. A place where a helicopter pilot could land. Kerner was bleeding heavily and could not move. Latino was walking around dazed. His helmet was gone and there was blood pouring down his face. Haggard took charge, told Private Thomas and Specialist Hanes to run over to the marijuana field they had passed on the way in.

"I knew what he was talking about but looking back, it's one of the things I think back on, and I [wonder] 'Did this really happen?'" Thomas recalled in 2019.

Hanes had the Minehound and was out in front. "We were so worried about our friends we don't use it. We just ran. Hanes is holding the Minehound, but we're running too fast to use it. We're just running and running on this path. We don't really realize anything, it's like nothing registers on us. Or I don't think of anything registering, until we made it to the marijuana field."

Thomas used the Defense Advanced GPS Receiver he carried, called a DAGR, to get coordinates. Ghariban said the coordinates were now good, that MEDEVAC could land in this open marijuana field. As Thomas and Hanes ran back, it dawned on Thomas what was going on. "We're in a very dangerous part of Afghanistan by ourselves and I'm just praying we don't get engaged. That we don't hit an IED, because we are dead if that happens. I would say we ran at least a quarter of a mile from where the IED blast happened, maybe more. Now we had to get back."

When they arrived back where the IED had gone off, the soldiers had tended to Kerner and now Haggard was carrying him on his back. He carried him like that all the way through the grape rows and to the marijuana field, where the helicopter could land.

"We popped smoke and a helicopter came," Haggard recalled. "Latino still had all that shrapnel in his face, but he was worried about his privates," Thomas said. They all wanted to have kids one day.

Back at Strong Point Payenzai, Twist learned the details of what happened and recalled feeling like "utter shit." There had been three IED hits in three weeks. Four soldiers were down. Latino and Kerner were seriously injured; Walley was missing limbs; Olivas was dead.

"Before I deployed, I read books about the war in Afghanistan," recalled Twist, citing "*Three Cups of Tea* and *Lone Survivor*." In one, an American mountaineer gets lost in Pakistan, gets saved by a friendly villager, and goes on to build sixty schools in Pakistan and Afghanistan in the first years of the war. In the other, a Navy SEAL in Afghanistan survives an ambush by the Taliban, escapes, and gets taken in by a friendly Afghan villager, who saves his life.

"What we were dealing with was none of that," said Twist. "They hated us. No one wanted us there. That was obvious."

PART IV

CHAPTER THIRTEEN

GETTING A PLATOON

When an American soldier is killed, wounded, or maimed on the battlefield, another soldier is brought in to take his or her place. This is the way of war. In June 2012, the 82nd Airborne needed to select a first lieutenant to replace Dominic Latino as platoon leader for First Platoon. When Clint Lorance, a young officer with no combat experience in Afghanistan, learned this position was going to him, he was excited. Among army officers, this is known as "getting a platoon." In the weeks prior, Lorance had complained to 2nd Lieutenant Katrina Lucas that it looked like he might return home to the United States without ever having left the tactical operations center at Forward Operating Base Pasab. That he might never get outside the wire or earn a combat infantry badge during his deployment to Afghanistan.

"He mentioned on several occasions that he was upset that he was the liaison officer rather than having an actual platoon like his first lieutenant peers," Lucas later told a military judge.

Katrina Lucas was an army intelligence officer who worked in the squadron's S2 shop. In April, after the battle space changed from Maiwand to Zhari, as part of the transition Lieutenant Lucas was assigned to be an assistant S2 for the squadron. Through the move to Zhari District, she acted as the intelligence liaison officer at Pasab, where she remained for three weeks until she moved to Siah Choy to be with the rest of the squadron. It was during this time at Pasab that she worked with Lieutenant Lorance.

"We interacted almost on a daily basis," Lucas told a military judge. "I would go into the brigade TOC and he would give me updates on what was happening with the squadron. In the evenings, I would attend the brigade BUB [battle update brief] with him. On occasion, we would have lunch together there at the FOB."

During their lunches, Lieutenant Lorance talked with Lieutenant Lucas about how he resented the other officers who made fun of him because he didn't have a Ranger tab. Katrina Lucas knew this was a real issue, she told investigators, and expressed empathy for Lorance regarding this dilemma.

"The majority of the combat arms lieutenants, if they don't have a Ranger tab or [if] they don't have a CIB [combat infantry badge] they typically feel they're falling behind their peers because they're not at the [same] standard as everybody else," Lucas said.

Clint Lorance had failed out of Ranger class at Fort Benning because of heatstroke. He could retry next year in cold weather, but that was too little too late. Deployments were tapering off. They both knew that, based on the current deployment timeline, the window of opportunity for a combat infantry badge was rapidly diminishing for Lieutenant Lorance. He had "only about five weeks of good deployment time [left]," Lucas agreed.

Then, on June 15, 2012, Lorance learned he was going to be "getting a platoon" after all. He shared how excited he was with Lieutenant Lucas. "Lorance was enthusiastic about taking over a platoon and expressed his desires to kill Taliban," Lucas later told criminal investigators.

In the S2 shop, Lt. Katrina Lucas worked with GEOINT products from the Persistent Ground Surveillance System (PGSS) team. Lucas's job involved "watching the battle space around Payenzai," and she clarified for a military judge what, exactly, that meant. "Our AO [area of operations] was a little unique to other AOs in the brigade in that a lot of [the Taliban] fighters [there] were not from the area. Once our unit moved in, a lot of the fighters dispersed to other AOs, specifically to the east and to the south. So they would exfil' and infil' into our AO by crossing over the Arghandab River to the south and then they would come into our AO from the east, typically from Pakistan." The fighters would go to locations at night, abandoned structures like the Grape Hut of Woe near where Specialist Olivas was hit, and the compound with bars on the windows, in Mullah Shin Gul, where Private Walley was hit.

One of Lucas's jobs was to send out to officers an intelligence product called a graphic intelligence summary, or a GRINTSUM, containing the most updated enemy activity available to a platoon leader. In June, shortly after Lorance learned he was getting a platoon, he emailed Lucas, requesting a GRINTSUM. Lorance had spent his entire deployment to date inside an air-conditioned tactical operations center watching live video streams from PGSS balloons and other overhead surveillance assets. Now he was being sent into active combat.

"He sent an email in response to one of the GRINTSUMs I sent out," Lucas told the judge. "And he mentioned that he liked the product and that there was a lot of good information in it. And at the bottom of the email he requested a 'be on the lookout list,'" also known as a BOLO. The BOLO is a list of wanted individuals whose identities have been culled from the biometrics-enabled watchlist, the BEWL, which is culled from the ABIS database.

In his email, Lorance also requested a "high-payoff targeting list," which Lieutenant Lucas said she provided to him. This list delineated everything in the area that could be "destroyed" by coalition forces: "personnel in the AO," meaning Taliban fighters and other insurgents, "or it could be weapons systems, vehicles; anything that would potentially be either taken or destroyed by coalition forces that will assist us in accomplishing our mission," Lucas said.

The list had a series of qualifying "identifiers," she clarified, meaning anything in the system that could be used to "either link back or relate to something that had happened in the past; [as in] a previous engagement." This link-back concept was central to the Joint Chiefs' pursuit of Identity Intelligence (I2), which involved "connecting individuals to other persons, places, events, or materials, [and] analyzing patterns of life." Lieutenant Lorance was interested in motorcycles, Lucas told the judge, because "two-wheeled motorcycles were considered an identifier." But in First Platoon's area of operations, two-wheeled motorcycles were a paradox, she said. They were two things at once.

"Red motorcycles would specifically be used in our area by IED emplacers," Lucas testified. Taliban "would drive the motorcycle up to a certain location, drop in an IED or activate an IED that was already in place, and then hop back on the motorcycle and drive away." The paradox was that a red motorcycle was also run of the mill: "[It] is a very commonly used vehicle both for the Taliban, and [also] for innocent civilians that are living and working in the [area] . . . Since the two most commonly used vehicles were red motorcycles and white Toyota Corollas, if we were preparing to go out on a mission, it would become kind of a joke that you were looking for those vehicles," Lucas told the judge.

One of the persons on the "high-payoff targeting list" who lived in and operated in this area was Abdul Ahad, the man whose "true identity" was about to become entwined with the story of First Platoon. He drove a red motorcycle too.

That a debate over this man's true identity would rise all the way to the desk of the president of the United States was almost inconceivable in June 2012. That this even happened has never been reported before.

In June 2012, according to Abdul Ahad's declassified biometric intelligence analysis report, his BIAR, seventeen months had passed since U.S. intelligence analysts had determined he was an IED manufacturer and Taliban commander. He'd been on the Joint Prioritized Effects List for many months. His status on the biometrics-enabled watchlist was "Tier 2: Question if Encountered." This is because in the context of the DoD's "Attack the Network" methodology Abdul Ahad could be more valuable alive than dead. "He is part of a larger [IED] network," analysts had determined.

Within Abdul Ahad's declassified BIAR, his IED identification link analysis chart reveals just how powerful a construct Identity Intelligence actually is. The scope and specificity regarding what the Defense Department knew about him, and his associates, is remarkable. Abdul Ahad had direct contact with at least fifteen Taliban commanders, all of whom were known to the Defense Department by name, six of whom were also on the JPEL, and therefore also marked for death. Abdul Ahad personally commanded at least twenty Taliban fighters, and was personally in charge of two Taliban financiers, brothers Mullah Agha and Abdul Agha, also IED manufacturers and emplacers. Abdul Ahad collected poppy-harvest taxes in the area; he planned attacks on combat outposts in Zhari District, including a rocket attack against FOB Pasab; appointed a Taliban judge to the judiciary; targeted Maiwand gubernatorial candidate Haji Qala Kahn for death; and made regular visits to Pakistan.

The personal details are similarly specific. Abdul Ahad took "small steps because of leg problems," had "two circular tattoos on [the] inside of [the] left wrist," wore "a black Dishdasha with a dark green vest and a Kandahar hat with no turban [and] a white Seiko watch." He "has two body guards, one with an 82mm mortar tube and one with a Chinese Kalashnikov," carries "a black radio, approximately 20 cm long, with a long antenna and four black knobs on the top." And he drives a "red motorcycle, new condition, Chinese-made, no windshield. Make: Honda."

But all of this information was mostly superfluous for a platoon leader guiding an infantry platoon on the ground. In Afghanistan, it was as if there were two different wars being fought at the same time. One war was digital. Driven by data. Seen from overhead, from the God's-Eye View. A war of geospatial intelligence, identity intelligence, and link analysis charts. The other war was human. Driven by flesh, blood, and bone. Experienced on the ground by infantry soldiers and Afghan villagers caught in this hopeless failure war. Lieutenant Lorance had, until now, experienced the war entirely in the first world; now he was heading into the second. No amount of data was going to bridge this gap. You can't shoot every man on a red motorcycle; that would be like killing every villager wearing a purple hat. That would be a war crime.

At FOB Pasab, as Clint Lorance prepared to head into the battle space to take the lead on First Platoon, the soldiers experienced another devastating loss, this one crushing in its wickedness, Private Twist recalled. It was June 23 and the soldiers were out on a presence patrol. They had just stopped in Sarenzai Village, which is approximately 600 meters northwest of the strongpoint.

"We were sitting down talking to the locals," remembered Specialist Haggard.

"We were there trying to figure out what they needed and what we could do to help them," Twist recalled.

"Sitting crisscross applesauce," Private Thomas remembered. "Sergeant First Class [Ayres] was talking through an interpreter."

Haggard lit a smoke. Thomas was sitting right next to Ayres because he was the soldier with the radio.

"I was just sitting there in my own world," remembered Thomas, "and the next thing you know, gunshots."

"Someone leaned around the corner and fucking shot Hanes," Haggard explained. "He was literally fifteen feet away."

"He sprayed bullets and disappeared," said Williams.

Carson pulled the pin on a grenade and threw it around the corner where the man had been. It blasted away part of a mud-brick wall. Thomas stood up and ran over to Hanes.

"Hanes was just laying there. He was just looking straight up at the sky. There was blood coming out of a bullet hole in his neck."

Specialist Joseph "Doc" Fjeldheim opened his kit and got the knife out for the tracheotomy.

"Doc's hand was shaking super bad," Thomas remembered.

"He had to do a trach on one of his best friends," Haggard recalled.

Sergeant Williams yelled an order at Doc Fjeldheim: "He's not your friend right now. He's a soldier you have to trach."

Private Thomas took Hanes's hand. "I knew he had some major problem, because he was just laying there. Doc cut in the throat for a trach to go in. We were holding him down, but he wasn't fighting back at all. And then I realized he's getting cut like this, and he's not moving." Which is when everyone realized right around the same time that the bullet had severed Hanes's spine. That he had been paralyzed.

"As Doc cut into him, I just see a big tear come out of his eye," said Thomas. "It rolled down his face and he wasn't saying nothing. His mouth was open. He had, I think, blood in his mouth and a tear coming down the side. And I just really felt for him. And I was just really, really concerned."

The soldier in charge of the litter that day forgot to bring it on the patrol. Sergeant Williams was enraged. Thomas and some other soldiers tried to make a stretcher using two rifles and a jacket. Sergeant Williams sent the Afghan National Army soldiers running into Sarenzai Village, and they came back with a door that they used as a stretcher. What happened after Hanes got shot was a blur for many of those involved. IEDs were evil. But to be sitting there, right in the middle of Sarenzai Village talking to a village elder, and to have someone lean around a corner and spray bullets, that was a new level of treachery and deceit.

Private Thomas doesn't remember calling in the coordinates for the helicopter. "It's very likely I did, but I really just don't remember," he said. "I was very, very upset."

The soldiers loaded Private Hanes onto the helicopter. The bird flew away.

When Clint Lorance left FOB Pasab on June 27, it was the first time he had gone outside the walls of a heavily fortified military base since he had arrived in Afghanistan the February before. Riding in a convoy of armored vehicles that included a road grader, he felt tense and fearful of an IED attack, he later told his lawyers, and that he "attempt[ed] to relax his body and his constant thoughts" but could not.

Two recent back-to-back suicide bombings in Kandahar City, approximately twenty miles to the west, had shaken him up, he said, and stayed in his head. Both of these attacks, which killed twenty-three people and injured fifty, used suicide bombers driving motorcycles. This was why Lorance had asked Lieutenant Lucas for the GRINTSUM, he later said. From intelligence reports he read at FOB Pasab, he knew there had not been any suicide attacks involving motorcycles in the 82nd Airborne's area of operations, in all of Zhari District. And there had been only one direct-fire incident involving a motorcycle in the area since May 26, a little more than a month ago. Still, he was focused on motorcycles, he said.

After Lorance's armored convoy arrived at Siah Choy, he checked in with Lieutenant Lucas to say hello. He told her how excited he was to finally be getting a platoon. He checked in with brigade commander Lieutenant Colonel Jeffrey L. Howard. He filled out paperwork for his change-of-duty station. On June 29, Lorance was driven in an armored convoy to Strong Point Ghariban. When he arrived there, he was shocked by the "craters, mortar holes and bullet holes," pockmarking the walls, he later told his lawyers.

At Ghariban, he met Captain Swanson, the officer to whom he would now report, and Sergeant Ayres, the soldier who would be his platoon sergeant at Payenzai. Captain Swanson asked Lieutenant Lorance if he wanted to meet the paratroopers, because everyone from First Platoon was presently at Ghariban. After Hanes had been shot in the neck and paralyzed, the platoon was pulled back from combat for several days of rest. A counselor had come down from Pasab and was available to any of the soldiers who wanted to talk to him. The platoon had sustained five casualties in less than four weeks.

Lorance wanted to meet the soldiers right away. He spoke to each of them individually, staying up late into the night, he said, and that his intention was to be a good leader. That was his plan. In the morning, armored vehicles would transport the platoon back to Strong Point Payenzai, where they would continue conducting presence patrols and capturing biometrics on villagers.

As for the young paratroopers of First Platoon, just when they thought there was no way things could get any worse, the deployment was about to take a sharp turn into a place no one could have forseen.

CHAPTER FOURTEEN

C-WIRE

The first of the armored vehicles shuttling First Platoon back to Strong Point Payenzai from Strong Point Ghariban arrived shortly after 7 A.M. on the morning of June 30. Private Twist was one of four soldiers in the MATV with Lieutenant Lorance. Pulling up to the entry control point, Twist saw an Afghan farmer standing there with a young child.

"He [had] a little tiny, tiny kid" with him, Twist later told a military judge. "Hard to determine someone's age in a war zone, but the kid had to have been maybe three or four because he was walking on his own."

Other than this father and child, the area around Strong Point Payenzai was a ghost town. Everyone was on edge because of what had happened to Specialist Hanes. Hidden IEDs had the terror of physical concealment. They could be blamed on an anonymous enemy, the Taliban. That a villager had just reached around a corner and shot Hanes in the throat while First Platoon's soldiers were sitting among them was perfidy. A new degree of deceitfulness and untrustworthiness. And it made Twist realize the feeling was mutual, he said. "Put yourself in their shoes."

To deal with the farmer and his young child, Lieutenant Lorance called for an interpreter. Private Twist, Specialist Bynes, and Specialist Matthew W. Rush pulled security, machine guns up at the ready. Out in the open like that, the soldiers were exposed. Afghan interpreter Fawad Elvis Omari came out from inside the strongpoint with an ANA soldier at his side. It was still early in the morning and the temperature had already reached 100 degrees. Today was Twist's birthday and he wondered to himself if he felt any different now that he was technically not a teenager anymore.

With Omari interpreting, the Afghan farmer asked Lieutenant Lorance if it was OK if he moved the army's concertina wire so he could access his grape field to farm. Lorance stood there in the hot sun, considering the man's request.

"Farmers were constantly coming to the [entry control point] to ask about moving the C-wire," Twist recalled. Once, a cow got stuck in the barbed wire. The animal was thrashing around in pain and had a bleeding leg. Twist and some other soldiers volunteered to help the farmer free his cow. The farmer was grateful, Twist recalled, and a few days later he brought some raw meat for the soldiers to share. They had a barbecue, which was technically not allowed but they did it anyway, and in 2019 this was a memory seared into Twist's brain. Thomas remembered it as "barbecuing raw meat."

The farmer stood there with his small child, waiting for an answer from Lieutenant Lorance.

"Move the C-wire, I'll have somebody kill you," Lorance told him. Then he nodded at Twist, Rush, and Bynes and the automatic weapons they held.

Twist watched as the interpreter hesitated to translate the lieutenant's threat. Rules of engagement forbade a soldier from conveying a threat. This farmer had been to the entry control point several times before, Twist would tell criminal investigators. There was only one farmer living in Payenzai Village whom the soldiers were on friendly terms with, and Twist was pretty sure the man standing in front them with a small child was him.

Lorance stared at the farmer. "We can make a deal," he proposed. "If you bring me IEDs, we can talk about moving the C-wire."

The interpreter translated what Lorance said.

The farmer said, "That's ridiculously dangerous," according to the interpreter. That trying to dismantle and pick up an IED without training was certain death. "I'm not going to do that," the farmer said.

"You can tell us where the IEDs are," Lorance countered. "You [can] bring us IEDs, or we'll have the ANA kill your family."

"He directly used those words," Specialist Rush told a military judge.

Specialist Bynes watched the farmer.

"I know of no IEDs," the farmer insisted.

Lieutenant Lorance pointed at the man's child. He said, "Do you want to see your child grow up?" Twist later told the judge. "He just pointed at the kid and [repeated], 'I'll have the ANA kill you.'"

The interpreter hesitated to translate this new lieutenant's series of threats.

"He didn't really want to say what Lieutenant Lorance was saying," Twist recalled.

Lorance told the farmer to be at the shura on Friday at nine A.M., and to bring twenty people with him. The Afghan farmer left with the small child.

That afternoon Lorance called the platoon together. From Ghariban, he had brought with him three items meant to outmaneuver the Taliban, he said. A small drone called a Raven, a Cerberus surveillance camera, and a bomb-dog team. Fitzgerald volunteered to test-fly the drone. Williams organized volunteers to get the camera up on a pole. Haggard and Carson were keeping watch in one of the guard towers when Lieutenant Lorance came to say hey.

"He looked out at the village," Haggard remembered. "He said if any of the villagers tried anything else, or if anyone else got hurt, he was going to fuck them up Nazi-style."

That night, Twist celebrated his twentieth birthday by smoking cigarettes and watching the movie *The Notebook* in the air-conditioned comfort of the MATV.

The next morning, the presence patrol began at 7:00 A.M. The platoon's mission that day was to walk north, patrol through the village, "to interdict any enemy forces they came into contact with," and return to the strongpoint. July 1, 2012, was the first day of a countrywide program called Afghan in the Lead. All across Afghanistan, ANA soldiers were now ordered to act as leaders, not followers. On presence patrols, this meant the Afghan soldiers would walk out in front of the American soldiers, not behind them as had been the case since the land-warfare branch of the Afghan Armed Forces was created in December 2002.

This day was even hotter than the day before, already 105 degrees. One after the next, the soldiers walked out of the entry control point and headed into the grape rows. The bomb dog could not climb or jump over the five-and-a-half-foot-tall walls inside the grape rows, and so the soldiers took turns passing the German shepherd up and over the vine-covered, insect-laden mud-brick walls. When they finally reached a village, Lorance held the American soldiers back but sent the Afghan soldiers ahead with the Minehound. He ordered them to conduct reconnaissance on one of the abandoned buildings there. After a few minutes, several of the Afghan soldiers returned.

Through the interpreter, the ANA soldiers told Lorance they'd found an IED in a doorway of the building. Sergeant Brian Peters went to look at it and discovered it was a metal doorjamb, not an IED, he told criminal investigators. That it was "a large metal object used to secure a metal gate to the ground," and definitely not a bomb. "There was hardened dirt covering the [metal object] and it did not have any secondary indicators to identify it as an IED," Peters clarified.

"Lorance told [me] he was going to blow it in place anyway," Peters said, and this upset him because not only was it unauthorized for a platoon leader to attempt to destroy an IED, but it was reckless and dangerous. If the object was an IED, protocol required an EOD technician disarm it. There was no EOD on hand, Peters told Lorance, not even at Strong Point Ghariban. And if it wasn't an IED, Peters said, ordering the platoon to stand around in the open to unnecessarily destroy a villager's property was foolhardy and illegal. Sergeant Peters told criminal investigators that "to confirm it was not an IED, he stepped on it to demonstrate." To show Lorance he was a fool.

Still, Lieutenant Lorance insisted on blowing up the doorjamb. He ordered the platoon sergeant, Ghulan Ali, to come with him. "[I was] grabbed by Lieutenant Lorance after [we] were [at] an appropriate stand-off distance . . . and [he] directed [me] to fire an RPG" at the metal doorjamb, Ali told investigators. An RPG, or rocket-propelled grenade, is a shoulder-fired antitank weapon equipped with a high-explosive warhead. Designed to destroy an armored tank, the RPG obliterated the entire doorway. After the smoke from the RPG cleared, Lorance informed the platoon it was time to head out. They would patrol back to the strongpoint on the road, Lorance said, not through the grape rows.

The bomb dog was exhausted and had lost its ability to sniff for bombs, so the minesweeper led the way. A soldier following the minesweeper trailed ten feet behind, marking safe passage with lines of baby powder. A few hundred meters before the platoon reached the strongpoint, insurgents opened fire on them from the tree line to the north. The soldiers took cover, laying down suppressive fire and taking turns running to the entry control point of the strongpoint. No one was hit.

Once inside the strongpoint, Herrmann requested permission to return fire with the Carl Gustaf recoilless rifle, a man-portable antitank weapon that surpasses an RPG in power. Lorance granted permission, and Herrmann fired at the tree line. The dog drank water. The soldiers who were off-duty went to sleep. Lorance wrote up a significant activity report noting the events from the patrol: the destruction of an IED and the indirect small-arms fire on exfil. With this report, Lorance met the threshold to earn his combat infantry badge.

That afternoon, around 5 P.M., Lieutenant Lorance went to Sergeant Williams with a question about Tower Two. As sergeant of the guard, Williams had been monitoring the radio traffic all morning. He was also educating himself on how to use the new Cerberus camera. It was a powerful piece of surveillance equipment, with night-vision capability and the capacity to zoom in on objects as far as two kilometers away. The Cerberus was now attached to a forty-foot pole at the center of the strongpoint, which meant whoever was using it could see around the garrison in a 360-degree view.

"Lorance came to me and said, 'Does Tower Two have a good vantage point of overseeing the village of Payenzai?'" Williams later told a military judge.

Williams said it did.

"How many people could be up there?" Lorance asked.

"Along the HESCO wall, you can get about two or three guys up there," Williams said.

Lorance asked Williams to locate the squad designated marksman and to have him meet Lorance up in Tower Two.

Williams located Staff Sergeant Christopher Murray, who was the squad leader for Specialist Rush, and told him Rush was needed in Tower Two. Williams continued educating himself on the new Cerberus camera.

Rush was sleeping when he was woken up by Murray.

"He told me that Lieutenant Lorance needed me up at the tower," Rush told a military judge.

Rush grabbed his Mk 14 Enhanced Battle Rifle, informally called an M14.

"I got my kit on, and I went up to the tower," Rush said.

Specialist Zettel, also a squad designated marksman, went too.

Inside Tower Two, Lorance briefed Rush about what he wanted him to do. "The reason we were there [Lorance said], the general mission, was to invoke the locals to come to the shura. The reasoning for us in the tower [he told me] was we were going to kind of recon by fire to get them to wonder why we were shooting at them so they would attend the shura." "Recon by fire" is short for reconnaissance by fire. Not a legal action, according to the rules of engagement.

"He directed me to get on the tower," Rush testified. "He said, he's going to designate targets for me [to] spark the locals' interest in why we were shooting at them." Rush had an M14 and 7.62mm rounds. Lorance had a pair of M22 binoculars. "It was daytime. It was very clear."

Lorance ordered Rush to "assume a prone firing position" so the two of them could work "as a shooter-spotter team."

Zettel never fired a shot. "I was pulling overwatch on [the] southern wall leading up to us so Rush and I didn't get shot [at]," Zettel clarified in an interview in 2019.

The distance from the strongpoint to the closest edge of Payenzai Village was approximately 450 feet. "We put a ghillie net over us, to conceal us and kind of keep us out of the sun," Rush explained, "ghillie" being a Gaelic term for camouflage.

Lorance ordered Rush to start shooting at the villagers who were walking around Payenzai—"civilians walking around the area," Rush clarified, which included old men, women, and children. Rush told the judge he remembered two targets specifically. One was a farmer tending to his crops.

"One was a civilian. He was 150 meters, 200 meters in front of the tower," Rush said. "And he's outside. And he's walking along a wall. And Lieutenant Lorance directed me to shoot approximately ten to twelve inches in front of him. And after I did that, [the man] stopped walking. He turned around and started walking the other way along the wall. And then [Lorance] told me to engage again, the same ten to twelve inches in front of him. And I did that again," Rush testified.

Every time the man stopped walking, Rush paused and then continued shooting "around" the man. He shot at his feet. He shot ten to twelve inches above the man's head. "It was kind of like I said, harassing fire." Rush was certain the farmer knew where the bullets were coming from because at one point, through his rifle scope, Rush saw the man look up at him in Tower Two.

Lieutenant Lorance seemed pleased. Next, he asked Rush to shoot into a group of children playing near a mud-brick wall.

"They were no more than 100 meters away," Rush told the judge. "[Lorance] designated me to do harassing fire [on] them. And that's when I said, 'You know, they're kids . . . I'm not doing that.' I [could] obviously see that they're kids without optics." Lorance again ordered him to shoot. Rush took aim and "shot at a door lock" instead.

Back in the headquarters tent, Sergeant Williams was working with the zoom on the Cerberus camera when he "began hearing fire." He heard two distinctly different kinds of fire, he later testified. Coming from the ANA tower, or Tower One, he heard gunfire that comes from a Remington 700. Coming from Tower Two, he heard the sound of 7.62mm rounds firing from an American M14. Williams walked over to the ANA tent and asked what the firing was all about. The Afghan soldiers told him they had received fire from Taliban hiding in an area to the north of Payenzai Village, and that they were returning fire.

Williams walked back to the strongpoint's makeshift operations center. Using the Cerberus camera, he scanned the ridge line to try to see who the Afghan soldiers were shooting at. The Americans were not in charge of the Afghan soldiers. Officially, they were partners. ANA commanders made their own decisions about who, or what, to engage.

"I couldn't see any people," Williams told the judge. "It looked like they were just shooting north of the village."

Next, Williams moved the camera lens so he could observe Tower Two.

"I saw our SDMs [squad designated marksmen] on the rooftop with Lieutenant Lorance. They were facing more towards the south."

Williams zoomed the camera out a little bit, in an attempt to see what they were shooting at. Williams told the military judge what he observed next.

"What I saw were women [and] children ducking for cover from bullets that were bouncing, impacting, off the walls next to them, and on the ground around them." Williams was alarmed, he said. It "was pretty obvious [the shooters in Tower Two] weren't trying to hit the civilians or hit the women and children," Williams recollected. It looked like the shooters "were trying to scare them," Williams told the judge, which was prohibited by rules of engagement.

As Williams considered what to do, Lieutenant Lorance came into the area where he was working. According to Williams, "He said, 'Ha-ha, that's, you know, it's funny watching those fuckers dance.'"

Sergeant Williams went to the back side of the tent, where

Sergeant Ayres was sleeping. Ayres was the highest-ranking NCO on base. "I woke him up to tell him he had to go up to Ghariban," Williams told the judge. "[I] informed him of what was going on with the shooting. [Ayres] was visibly upset and ran outside to talk to Lieutenant Lorance."

Sergeant Williams was in a bind. "I did not want to get in between my platoon leader [Lorance] and my platoon sergeant [Ayres]."

Five minutes passed.

Williams could still hear shooting coming from Tower Two. "[Lorance] had told [Zettel and Rush] to stay up there for a couple more minutes and 'dump another mag' into the village," Williams told the judge.

Several more minutes passed.

"Lieutenant Lorance came [back] in and he told me to report up [the chain of command] that they had taken potshots. That the platoon's SDMs had received potshots from the village," Williams told the judge. "I told him that I wouldn't do that, I wouldn't report that up. Because it's a false report." No villagers had fired into the strongpoint from Payenzai Village. Lorance was asking Williams to lie.

Ayres came into the operations center. Lieutenant Lorance and Sergeant Ayres began arguing. Williams heard Lorance tell Ayres, "If I don't have the support of my NCOs then I'll fucking do it myself." He stormed out.

Later, Lorance came back and tried to explain to Sergeant Williams why he did what he did. "He told me the reason he was doing this was to get people to come to his shura that was supposed to be that Friday. [That] he would make them fucking come one way or the other. He said he didn't really care about upsetting them too much because he fucking hated them."

Williams stood staring at his commanding officer. He did not know how to respond. Lorance continued talking. "Lorance said that he was in love [with] our platoon and had been for a while,"

Sergeant Williams told the judge. "That he had been monitoring us since the beginning of deployment and [he] really loved our platoon. And he didn't [want to] see any more of our guys get hurt."

That evening, Sergeant Ayres made a pivotal decision. He did not write up a report indicating that Lieutenant Lorance had asked another soldier to lie to command.

"We wanted to give him a chance," Williams said. "The lieutenant was new. It was his first day in the AO. We all knew he'd spent the deployment in the TOC [the tactical operations center]. Before that, he was military police in Iraq. Who joins the U.S. Army to police other soldiers? Fucking weird. Never mind. He hadn't been at Payenzai before. We thought, OK, now he'll calm down."

All day on July 1, Captain Swanson had been with squadron commander Lieutenant Colonel Howard at Pasab. "When I got back [to Ghariban], I read the report of the [significant activity] just because it seemed to lack a little detail. It did not make complete sense," Swanson told the judge, referring to the doorjamb Lorance had written up as being a twenty- to thirty-pound IED. "I was also concerned about the amount of force that was reported," Swanson told the judge, referring to the Carl Gustaf rifle that Sergeant Herrmann fired after the platoon returned to the strongpoint.

At the end of each day, as was protocol, Swanson conducted an operations security meeting with his leaders. On July 1, this meeting took place at 2100 military time, which is 9:00 P.M. Swanson briefed Lorance on the use of force and the principle of proportionality in warfare. "Proportionality is using the least amount of deadly force required in response to a hostile act. It's a general guiding principle. I also talked about positive identification," Swanson told the judge. Positive identification is a key concept because in the U.S. Army, you must positively identify someone as an enemy engaging in a hostile act before you fire your weapon at them.

In the U.S. Army, you fire your weapon for one reason and one reason only: you shoot to kill.

"I made a point to bring up the importance of positive identification as a requirement, among other things, for using lethal force against somebody. I kind of wrapped it all up by saying without having a proportional response, and without having PID [positive identification], I think my exact words were 'You're going to kill somebody who shouldn't be killed.'"

CHAPTER FIFTEEN

JULY 2, 2012

The morning of July 2, 2012, began like other mornings, with the soldiers up at 4:30 A.M. to get ready for the pre-mission brief, followed by a dismounted presence patrol. Lieutenant Lorance wanted to leave the strongpoint early to avoid getting pummeled by the crushing midday 112-degree heat. Today was going to be even hotter than it had been the day before. In the early-morning darkness, Specialist Fitzgerald had figured out the Puma had broken and the handheld surveillance drone wouldn't fly. The AeroVironment RQ-20 Puma was a high-tech warfighting tool carrying an array of electro-optical sensors. It ran on batteries, was hand-launched, and cost $250,000 per drone. When it was missing a part, as it was on this day, its capabilities were reduced to zero. Lorance insisted on getting it fixed immediately, before they headed out into Sarenzai Village, and threatened to take Fitzgerald's rank away if he didn't. Lorance grew angry, upset by the Puma's failure.

Things were tense. The last time the platoon had patrolled into

Sarenzai Village was when Hanes had been shot in the neck by someone hiding behind a wall who then disappeared. During this July 2 pre-mission brief, Lorance wrote a message on the whiteboard. It read: "Deny Enemy Sanctuary and Get Payback for Hanes." Lieutenant Lorance did not know Matthew Hanes, or any of the other soldiers who had been killed or maimed by IEDs, but he was familiar with what had happened to each of them from his work in the TOC at Pasab.

"Revenge is not part of what you learn in the army," said Private Twist. "We had pre-deployment classes that dealt with revenge because, let's face it, that's what some people think war is [about]." Twist remembered a soldier from that class. "We went around saying why we joined the army. One guy said he joined to get payback for his brother who got killed in Iraq." The instructor spent a long time talking about how that was fine to feel but not fine to act on. That "revenge was not what the army was about."

Lorance made an announcement. He told the platoon there was new intelligence from Colonel Mennes, the brigade commander.

"We were told by Lieutenant Lorance that we were to engage any two-wheeled motorcycles on sight," Fitzgerald later told a military judge. "Somebody objected. And [Lorance] told us that he had received intelligence in the area that anybody on a two-wheeled motorcycle in the area was Taliban, and was to just be engaged on sight."

Sergeant Williams told the judge he was surprised by what Lorance said. "He said . . . any two-wheeled motorcycle was to be shot at. And any three-wheeled motorcycle, if it had children on the back, it was to have special consideration. It was to be interdicted."

What Lorance was saying did not make sense to any of the platoon members interviewed for this book.

"All of the members of the platoon that were going on [the] mission that day, we all looked at each other," Fitzgerald told the military judge. "We all knew something was wrong. We knew what our rules of engagement were prior, and we had not received any changes prior to this day."

As the platoon prepared to head out, Lorance told the soldiers that they were going to get revenge in a shock-and-awe campaign through the village.

"Shock-and-awe" were the exact words that he used, Private Skelton told the military judge.

Shock and awe is a military strategy that became generally known to the public after President George Bush used the phrase to describe the opening salvo in the 2003 Iraq war. The tactic uses a "spectacular display of force and power" designed to paralyze the enemy's perception of the battlefield and destroy their will to fight on.

It was time to head out. At the entry control point, assault squad leader Corporal Jarred Ruhl got the men into a staggered file formation. Lorance ordered the two gun trucks to move into position, to act as the support by fire element to keep watch over the day's patrol. Inside the gun truck closest to the village were Specialist Reynoso, Private Shilo, and Specialist Frace, from weapons squad.

The Afghan National Army soldiers headed out from the strongpoint first. They crossed the dirt road, ducked under the C-wire, and began moving into the vegetation inside the grape rows. The American soldiers followed next, one at a time, spaced ten to fifteen feet apart, with the Afghan soldiers in the lead.

Two miles to the west, inside the command and control shelter at Siah Choy, PGSS aerostat operator Kevin H was seated at his workstation when radio operator Sergeant Watson, from fires group, came in with a request. Kevin H needed to swing his PGSS over to the east, Sergeant Watson said. First Platoon was leaving Strong Point Payenzai, and Kevin H needed to identify a possible threat.

Earlier that morning, at 6:19, the joint operations center at Pasab reported hearing about a "pending ambush." Insurgents were seen gathering roughly 900 feet to the north of the strongpoint, just beyond Sarenzai Village. According to official records, the operations center had "received ICOM chatter indicating a pending ambush on a combined patrol moving toward an INS [insurgent] position." A few minutes after this report came in, the joint operations center sent its message to Siah Choy, requesting that a surveillance asset be brought overhead of the strongpoint to cover down on the platoon.

Kevin H moved the aerostat down from 2,100 feet to 1,600 feet. He positioned the MX-15 camera on his surveillance objective: Strong Point Payenzai.

"Within seconds," Kevin H stated in a sworn affidavit, "I saw three figures with AK-47 rifles, an ICOM radio, and binoculars. Most likely Taliban . . . shadowing the patrol." Kevin H confirmed with Sergeant Watson that the insurgents, moving on foot, were approximately 900 feet to the north of the entrance of the strongpoint. At this distance, the insurgents with the AK-47 rifles were too far away from the platoon to be considered an immediate threat. Kevin H watched the soldiers move forward, slowly making headway through the grape rows. After a few minutes, he was told he needed to move his surveillance camera elsewhere, to identify another threat facing another platoon, in another village down the road.

At Strong Point Payenzai, there was an unforeseen problem at the gate. Several of the soldiers had already exited the strongpoint, but Private Skelton, Specialist Haggard, and Specialist Fitzgerald were still there, standing next to Lieutenant Lorance. Three Afghan villagers on a red motorcycle had just driven up to the entry control point and were demanding to speak with the officer in charge. An Afghan interpreter translated the villagers' concerns.

"They began complaining about having received some fire into their village from one of our guard towers," Fitzgerald told the military judge. The harassing fire Lorance had ordered Specialist Rush to engage in the evening before, from Tower Two.

Skelton, Haggard, and Fitzgerald stood beside Lieutenant Lorance as the Afghan men verbalized their complaint. Skelton, Haggard, and Fitzgerald each individually recognized the men on the motorcycle as local villagers. One of the men was the local tribal leader. He had a long white beard and appeared to be about sixty years old.

"They were known to me from previous missions into the village of Payenzai," Skelton later told the military judge. "They had been biometrically enrolled by me on more than one occasion. They were usually working the fields around the area."

Lieutenant Lorance told the men to leave. They refused.

"They wanted to complain about the fact that they were being shot at the night before," Skelton told the judge.

Lorance told the Afghan men on the red motorcycle that if they had a problem with anything, they needed to come to the weekly meeting where grievances were addressed.

"Come to my Friday shura," Lorance told them.

But the men refused to leave. Lorance began aggressively shouting at the men. He racked his M4 and began counting down: "Five . . . four . . . three . . ."

As Lieutenant Lorance began counting down like that, the three Afghan men climbed back onto their red motorcycle and drove away.

Inside the strongpoint, through the Cerberus camera, Sergeant Williams had been watching this altercation at the entry control point.

"I didn't hear the conversation between the villagers and Lieutenant Lorance. I could only view it on the camera," Williams told the judge. "Lieutenant Lorance called me up on the radio and he said, 'Hey, don't be surprised if these guys come back in five minutes when we're gone. And when they do come back, you tell them to leave, and to come back fucking Friday. And if they don't, and if they won't, then shoot them in the fucking face.'"

Approximately twelve minutes passed. Pfc. James Skelton was in the middle of a farmer's field, on top of a grape row, approximately 200 feet from the entrance to the strongpoint, when he saw three men on a motorcycle, driving fast toward the platoon. They were driving "between 30, 35, 40 miles an hour," Skelton later told the military judge. "The vehicle was driving fast down Route Chilliwack," he said. He told criminal investigators that he feared for his life. "The first thought I had was, it could be a drive-by shooting; it could be a drive-by grenade throw; it could be a vehicle-borne IED," he told the military judge.

Within seconds, following the chain of command, Private Skelton asked Lieutenant Lorance permission to open fire on the three Afghan men riding the motorcycle. Skelton did not yet realize these were the same men who had been at the entry control point just a few minutes before. Lorance was standing behind Skelton in the order of formation, down in a culvert inside the grape rows. His view of the motorcycle and its three riders was obscured by a mud-brick wall within the grape rows. Lieutenant Lorance granted Private Skelton permission to open fire on the men on the motorcycle.

"I fired two rounds, missing both," Skelton testified in military court. Later, he told criminal investigators that he had "aimed to hit center mass of the driver of the motorcycle. [I] did not fire a warning shot and did not intend to miss." He said that he "dropped his weapon about a week prior, and had not zeroed it since."

Fitzgerald was the soldier standing closest to Skelton when Skelton first opened fire on the three riders. Standing six feet three inches tall, Fitzgerald was taller than the mud-brick walls. "[I] had clearance over the grape row," he said. "I was taller than the vegetation was."

Fitzgerald heard Skelton fire rounds from the M4 he carried. He immediately turned and pulled security, as he had been trained.

"I needed to make sure that we weren't going to receive any fire from the wood line to our north," Fitzgerald told the judge, referring to the abandoned village of Mullah Shin Gul. "We had taken fire from [there] in the past."

Fitzgerald was 528 feet away from the three Afghan villagers on the motorcycle. As the designated marksman for the platoon, Fitzgerald carried an M14 Enhanced Battle Rifle outfitted with a 10X optic scope. The telescopic sight made distant objects appear magnified. Looking at the scene through the glass meant Fitzgerald saw the men with clarity. He'd seen the Afghan men riding together on the motorcycle. To Fitzgerald's eye, they were "not traveling very fast." He had heard, and seen, Private Skelton open fire on the men. He'd seen Skelton miss.

"I witnessed the motorcycle come to a full stop, and the three individuals got off the motorcycle," Fitzgerald testified. Through his scope he watched as one of the men put the kickstand down with his foot. Fitzgerald could see clearly that the men had their hands up and were waving to the Afghan National Army soldiers standing just ahead of him. There was an exchange of words.

"The ANA started telling them to go back, waving them to return towards the motorcycle," Fitzgerald testified.

To Fitzgerald's eye, the men who had just been on the motorcycle were now conveying the universal two-handed gesture for *What is going on?*

Then, over a radio, Fitzgerald heard Lorance issue a fateful command. "Lieutenant Lorance ordered us to open fire" on the three Afghan men, Fitzgerald told the judge.

Fitzgerald had the three Afghan men in his sights, he told the judge, but did not fire his weapon because he could see clearly that they had their hands up and were not a threat.

Back at Strong Point Payenzai, Sergeant Williams used the Cerberus camera to watch what was happening in the battle space, as it was happening, in real time. Williams was the highest-ranking noncommissioned officer at the strongpoint. Using the camera's powerful zoom lens, Williams had been panning back and forth between two positions: the soldiers in the grape rows and the gun truck on the road. When Williams saw the three Afghan men on the red motorcycle come into view, he immediately notified Private Shilo, the M240 gunner in the gun truck, as was protocol.

"Sergeant Williams, who was monitoring the raid camera, he called up to our truck to notify," Shilo told the judge. "He notified us that there was three PAX [passengers] on a motorcycle moving somewhat towards us on the route, from the north." Because he was the M240 gunner in the armored vehicle, Shilo's head and upper body would sometimes be out in the open air. "In the middle of the vehicle, there's a hatch with a turret," Shilo clarified for the judge. "And mounted on the turret was a M240 Bravo. And what I did was, I stood up in the turret and just scanned the area."

Shilo looked around. "When I first looked for them, I couldn't find [the three villagers]," he said. "After about a couple minutes, they started to come out of, like, a little tree line on a small cow path that they were moving along, [near] a wall which bordered the outskirts of the town." Shilo continued to watch the men. "[They were] looking around, trying to figure out what happened," Shilo told the judge. "Not necessarily like they were scared or like they were nervous or anything, but just kind of trying to figure out what had happened. If the shots [fired by Skelton] were directed at them or not."

With his head and shoulders still outside the turret hatch, Shilo kept his eyes on the motorcycle riders. Down below, Reynoso sat balanced on the gun truck's passenger seat. Frace was in the driver's seat. Reynoso couldn't see outside. His view of the riders was obscured. Reynoso was talking to Lieutenant Lorance over the radio. As the designated truck commander, it was Reynoso's job to relay orders to Shilo as the orders came down the chain of command.

Back at the strongpoint, through the Cerberus camera, Sergeant Williams watched the actions of the motorcycle riders. "They began pointing at our formation," he told the judge. "They obviously knew we were coming [through the grape rows]."

Then, a watershed realization for Williams: "I recognized the older gentleman [on the motorcycle] as the village elder, [a man] who we called the village elder because he was always the one that would come out and talk to us," Williams later testified. The tribal leader with the white beard.

Inside the grape row, Corporal Ruhl was two or three men away from Lieutenant Lorance, directly in front of Private Skelton in the grape rows. "After that happened, after [Skelton] missed the shots," Ruhl testified, "Lieutenant Lorance contacted the gun truck [over communications] that was to the northeast of where we were at in the [staggered file] movement and asked them [in the gun truck] if they had eyes on the bike." Still down in a gully, in the grape rows, Lorance could not see the three motorcycle riders. He did not know they had dismounted from the motorcycle, put the kickstand down, and were now attempting to communicate with the Afghan soldiers verbally and also with their hands and arms.

Half-inside, half-outside the gun truck, Private Shilo kept his eyes focused on the motorcycle riders. "They stayed there for a couple—about two to three more—minutes and they [were] still wandering around, and as they got onto the bike, that's when I was given the order, if I still had eyes on the men, to engage. I said, 'I did.'"

Private Shilo had been given a direct order by his commanding officer, Lieutenant Lorance, to kill the riders.

"As soon as I was told to engage," recalled Shilo, "I cocked back the 240 and began to fire. When I first shot, it jammed, so I re-cocked it back, performed a functions check, and then I engaged and I hit the first individual and, you know, he kind of got up, staggered, and fell."

Private Shilo fired again.

"The second [man] went to go look at him [the dead man] and at that time, that's when I hit him." Two of the three riders were now lying on the ground. The third rider stood upright. He was dressed in all white. He turned and sprinted away.

Through the scope on his rifle, Fitzgerald kept his eyes on the men as they were being killed.

"The gun truck opened," he told the judge. "I heard a burst and I watched one of the individuals fall. The second one, just a moment later, went down."

Fitzgerald watched the third man, the uninjured rider dressed all in white, break into a run and disappear into the village. Fitzgerald moved his eyes back to the two men lying prone in the road. "At that point they no longer moved," Fitzgerald told the judge. "They just laid there, still."

Fitzgerald stared at the bodies splayed out on the dirt road in a crumpled heap. There was a dark pool of blood spilling out from the body of one of the men and it was growing bigger by the minute, staining the tan-colored earth. Then Fitzgerald moved his eyes to the tree line to the north. He was down on one knee, all six feet, three inches, 170 pounds of him. When Fitzgerald was a kid, people made fun of his anatomy, of his height, and called him the skinny stick. He joined the army to prove them wrong. He was a dedicated soldier who knew his job. Now here he was in Afghanistan, down on one knee, pulling security to make sure no one else got killed. He had just watched two men with their hands up get shot and die. He watched the third man run away.

Everything here was hell. Afghanistan was where everything

that meant anything to you got destroyed. Now it was all spiraling out of control.

The three Afghan villagers riding on the red motorcycle were a father, his brother, and one of his sons. The father, the village elder with the long white beard, went by the name Haji Mohammed Aslam. He and one of his sons were shot and killed. The son went by the name Ghamai Abdul Haq. The village elder's brother was the rider who ran away, the man dressed in all white, and who went by the name Haji Karimullah.

— Haji Mohammed Aslam
— Ghamai Abdul Haq
— Haji Karimullah

Outside of the people living in the village, and the members of the men's tribe, no one knows very much else about the identities of these three men. But the record indicates that these were their names.

The Defense Department does not trust names. "Terrorists hide behind their anonymity," Vice Admiral Robert Harward wrote in the *Commander's Guide to Biometrics in Afghanistan*. This is why the military's biometric database, the Automated Biometric Identification System (ABIS), had been created in the first place. To the U.S. Department of Defense, it was a man's alphanumeric biometric identification, his BID, that mattered most. That was what the Manhattan Project–style, billion-dollar biometrics program was all about.

CHAPTER SIXTEEN

THE MAN IN WHITE

Immediately after the two Afghan villagers were killed, Private Thomas used his radio to notify the tactical operations center at Siah Choy of the engagement against the three motorcycle riders. Thomas had been standing next to Lieutenant Lorance when Lorance ordered the gun truck to kill the motorcycle riders.

"A small moment after that," Thomas told the judge, "CCA [close combat attack] helicopters checked on station. They sent me comms over the radio saying that they're in our area and that we could use them for what we needed at that moment."

Private Thomas relayed this information to Lieutenant Lorance, who was standing right beside him.

"First Lieutenant Lorance told me . . . he pointed at the motorcycle that we just shot at and he was referring to the individual in white that [just] ran away from us, and he said, 'Thomas, I want that guy dead by CCA. Time: Now!'"

One of the CCA helicopter pilots asked Private Thomas for a description of the man Lieutenant Lorance wanted killed.

"The only description I could give them," Thomas testified, "was that he was in all white."

Private Thomas turned back to Lorance. "I asked First Lieutenant Lorance what a better description of the man would be, and First Lieutenant Lorance's reply to me was, 'I don't want you to give a good description of the person.' He said, 'You never want to give CCA a good description of the person, because then that makes them seem innocent.'"

One of the pilots in the attack helicopter was 1st Lieutenant Catherine McNair, a member of the U.S. Army's Second Squadron, 6th Cavalry Regiment, 25th Combat Aviation Brigade. The air cavalry's job was to conduct reconnaissance, security, and attack missions for units across southern, Afghanistan. McNair piloted the helicopter over the area to photograph the engagement site and conduct general reconnaissance. She and copilot Chief Warrant Officer 2 Matthew Pierson would stay on-site for the next forty minutes, McNair later testified.

"It was requested that we look for an individual meeting a certain description and that he had been deemed hostile and if we were to find him, we'd be clear to engage." The certain individual, McNair testified, was the third motorcycle rider, described to her as wearing "all white." "Clear to engage" means clear to kill.

As the helicopter pilots continued their search for the man in white from above, they could see a mob of villagers gathering on the ground below. Through comms, McNair relayed to the chain of command that she had identified "a larger group of adult males massing and then departing the objective area." The group, McNair said, was made up of "seven or eight military-age males," and were gathering just north of the village, in a manner that the Defense Department categorized as "a common TTP [tactics, techniques, procedures] for insurgents."

Some of these men were on foot, McNair reported; others were on motorcycles. The man dressed in white was not among them, McNair testified. No one in the group appeared to be carrying weapons or ICOMs. According to the rules of engagement, without weapons or ICOMs, these men could not be engaged. McNair dropped smoke on this group of military-age males and watched as the men split up and disappeared into the crowd.

At Ghariban, Captain Swanson was notified of the killings.

"The contact was over at that point, and so my immediate concern, then, was to do the BDA, the battle damage assessment," Swanson told the military judge, "enrolling the casualties into our biometric identification systems to match it with the catalog of people that maybe we have come into contact with." This action was critical, Captain Swanson told the judge, and needed to happen now. "Aside from establishing security, frankly, conducting the BDA was the next most important thing within the brigade to execute."

A BDA was the platoon's way of checking its work, Swanson said. The BDA was about establishing facts. "You want to be the first with the truth. And in order to do that, you've got to go and conduct the battle damage assessment."

The BDA, and the biometric information it includes, allowed a company commander like Captain Swanson to distinguish whether his platoon had killed innocent farmers or Taliban insurgents. To determine truthfully whether the deaths were civilian casualties, CIVCAS, or enemy killed in action, EKIA. This situation was complex. The person who had initially called out the threat, and had opened fire on the three motorcycle riders but missed, was Private Skelton, First Platoon's COIST member. And at first, Private Thomas thought the ANA had opened fire, too.

On the ground, the situation at the entrance to Sarenzai Village was turning into mayhem. Lieutenant Lorance ordered the ANA soldiers, the weapons squad, and part of third squad to push into the village and begin doing a house-to-house search for the man in all white.

"Myself and Private [Paul] Copeland were set on security at the rear," recalled Fitzgerald. "Closest to the motorcycle and the individuals that had been shot. We were told actually specifically to shoot anybody that approached the motorcycle or the bodies," Fitzgerald testified. A crowd of people was moving toward them.

"The children and the women were crying, screaming, and crying. They were carrying large sheets," Private Skelton told the judge.

The villagers' sheets obscured the soldiers' line of sight and added to the pandemonium. More people were gathering. Wailing for the dead. The paratroopers were shouting above the chaos, trying to keep the situation from spiraling further out of control.

"We were approached by an adult male, an adult female, and a couple of children," Fitzgerald told the judge. "They were crying. They were upset. They said that they were relatives [of the men killed]." Through an interpreter, they asked to retrieve the bodies of their family members.

"Lieutenant Lorance told them that they could not go and retrieve the bodies, they couldn't go near them, and that we would shoot them if they did so," Fitzgerald testified. "Lieutenant Lorance became irate, began yelling at them. He was actually talking through the interpreter, but he was talking to the locals, and he told them specifically, 'Shut the fuck up or I'll shoot you too.'"

Overhead, the attack helicopter continued to circle. On the ground, Fitzgerald could see the motorcycle. It was upright and the kickstand was down. Over comms, Lorance spoke directly to Private Shilo, in the gun truck.

"Do you see the motorcycle?" Lorance asked Shilo.

Shilo said, "Yes, I still have eyes on it."

Lorance ordered Shilo to shoot the motorcycle. "To disable it."

As Shilo prepared to fire, he witnessed a young boy emerge from the crowd of villagers and begin walking toward the motorcycle. The lieutenant had given him a direct order to shoot at the motorcycle. This time, he did not shoot. At the trial, the defense attorney asked Shilo why, in that moment, he refused an order from a superior officer in a combat zone.

"I wasn't going to shoot a twelve-year-old boy, sir," Shilo said. He watched the boy retrieve the motorcycle and disappear into the village.

Sergeant Ayres hurried over to Lieutenant Lorance and told him there was a threatening situation developing to the north.

"We were intercepting ICOM chatter," Ayres testified.

The chatter was coming from the grape rows near the abandoned village of Mullah Shin Gul. The Taliban might be setting up for an ambush, Ayres said. Sergeant Herrmann took Specialist Haggard, Private Carson, and Specialist Cole Rivera from weapons squad with him to set up an overwatch position on a rooftop. The soldiers walked west down the cow path until they reached an abandoned building they could set up overwatch on. Haggard and Carson climbed up onto the roof, passing weapons fast. They quickly set up a machine-gun position and scanned the area to the north. Through a laser rangefinder, Carson observed a military-age male talking into a radio.

"The ICOM looked like an old walkie-talkie with a silver antenna ten to twelve inches long," he told investigators. "[I] spent three to five minutes observing."

Haggard scanned the area too. "[I] saw a grape hut [that] had six males around it," Haggard told investigators, "observed the male talk on the ICOM radio and point in [our] direction . . . I saw a man about 360 meters out with my naked eye and then I verified it while looking down my optic."

According to declassified communication intercepts, called Wolfhound chatter, one of the Afghan men in this group spoke into an ICOM radio. "There are Americans on the roof. We want to do something to them," he said.

Haggard scanned the area for movement. "In the third or fourth grape row, there was a man that popped out," near a pump house. He was talking into an ICOM. "There's a bunch of ICOM chatter coming [through comms]. You could tell they were looking at us and watching us. They were watching our element push through, and they were setting up to attack."

Carson saw it, and called it out too. As the highest-ranking noncommissioned officer in the squad, Sergeant Herrmann ordered the men to open fire.

"I directed it," Herrmann told investigators.

"Sergeant Herrmann said for the gunner to eliminate him," Haggard recalled. "I was the gunner. I put four bursts into him before he went down."

Haggard continued scanning the area. There were at least two other men in the vicinity, hiding in the grape rows. He watched the head of one man pop up, then disappear. A man ran over to the man Haggard had just shot, now lying dead in the grape row. "He grabbed the ICOM and tried to run away. [Sergeant] Herrmann shot at him first, missed. I shot. He went down. Then you could see . . . everybody else scatter."

This second man had been hit in the arm. Wounded, but still alive. Haggard looked at the man he'd just shot with four bursts. The man showed no signs of life. "I watched four little kids and a woman with a wheelbarrow go pick him up" and cart his body away, Haggard told me in 2019.

Haggard and Carson were exposed on the roof. Rivera and

Herrmann kept overwatch as they climbed back down. The soldiers flex-cuffed the detainee who had been shot in the arm. He was their prisoner now. He said his name was Mohammad Rahim.

Back at the entrance of Sarenzai Village, Lieutenant Lorance, Private Skelton, and Specialist Frace waited along the roadside by a tall tree. Fitzgerald and Copeland pulled security, standing closest to the bodies of the two villagers who had just been shot.

Fitzgerald kept watch over the bodies, also scanning the wood line to the north. Just a few feet over from him, he told the judge, Private Skelton stood with Lieutenant Lorance, arguing hard beside the tree. Privates don't argue with officers. This defies chain of command. Fitzgerald knew something unusual was going on. On the ground ahead of him the bodies lay exposed in the sun. The earth beneath his boots was parched and splitting. The tree looked dead but for a small patch of leaves.

Private Skelton was arguing with Lieutenant Lorance about biometrics.

"I'm the one who does that, sir," Skelton told Lorance.

Private Skelton insisted that he needed to get the biometrics on the Afghan men who had just been killed. When someone got killed, it was Skelton's job as the platoon's COIST to capture biometrics from the body, usually the iris scans, he said.

Lorance told Skelton he did not want Skelton to do a BDA. He gave him a direct order to stay back.

"We don't necessarily want to do a BDA, because some people aren't necessarily enemies," Lorance said. Private Thomas heard it. Specialist Fitzgerald did too.

Fitzgerald watched as the argument between the private and the lieutenant escalated.

"Don't you want me to go SEEK and Expray?" Skelton asked Lieutenant Lorance for a second time.

"No," Lieutenant Lorance said. "And I'll tell you why in a minute."

Before explaining, Lorance called for Private 1st Class Erik Wingo and Spc. Reyler Leon to come over to him. Wingo was new to the platoon, having replaced Hanes after Hanes got shot. Lorance ordered the two soldiers to search the bodies of the two Afghan men the old-school way, using their hands. Lorance did not want Skelton or anyone else to take biometrics with the SEEK.

"You would not necessarily like what you see," Lorance said.

Wingo went first, sweeping the ground in front of him with the Minehound. The road was crooked and made of hard-packed earth.

"Myself and Specialist Leon approached down the path from the west," Wingo told the military judge. Leon flipped over the body of one of the dead men and examined him from the front.

"We found one gunshot wound to the chest," Leon testified. "I checked for any signs of life: by carotid arteries, wrist, and any respirations of any type, and I found nothing."

The soldiers searched the body of the second man.

"We [located] a gunshot wound to the neck. We also checked his breathing, pulse. To confirm that he was dead."

"On the bodies," Wingo testified, "we found an identification card; a tightly rolled-up piece of paper; three cucumbers; several ink pens; a small pair of scissors; a small gourd-like object. It was veneered. It had a finish on it. It was probably something you could buy at [a] market."

As for weapons, "There were no weapons," Wingo told the judge. "There were no cell phones, no ICOM radios, and no weapon[s]."

Skelton and Lorance stood beside the tree, watching from approximately 100 feet back. As they stood there, something alarming happened.

"The local national that evaded the fire from the machine gun, on the gun truck. [He came] back," Skelton told the military judge.

"He was wearing the same exact garments that he was wearing. Clothing," Skelton clarified.

The man in all white. The man the platoon was doing a house-to-house search for in the village. The man the attack helicopter was flying overhead, in circles, looking for—to kill on the lieutenant's orders. He had walked up and was now standing next to Skelton. The man in white had several family members with him, Skelton said. They were crying and giving information through the Afghan interpreter and asking him questions.

"The interpreter was with me," Skelton told the judge. "And they were identifying the deceased individuals as their family members. And they were giving their names, [and] I recall[ed] meeting [them] before."

At the trial, even the prosecutor was caught off-guard.

"Now, you've never said that before either, have you?" the prosecutor asked Skelton. "That [this man] identified himself as the guy that ran away?"

"When I saw the individual had run [away] and come back," Skelton said, "I [realized] he was a local villager. I saw the family, which I had seen and interacted with before."

The prosecutor asked again, just to make sure. "And you did not tell anyone that? Because the patrol says they never found that person."

"The patrol went through the village, and they wouldn't have found him [because] he was with me," Skelton testified.

The third man on the motorcycle, Haji Karimullah, was standing beside Skelton—Skelton being the soldier who, just minutes earlier, had fired on this man, his brother, and his nephew, with the intention to kill them.

Leon and Wingo were done with the old-school BDA, one they performed with their hands, not with an automated machine. Now it was time to head back to Strong Point Payenzai.

"The locals started to move down the road towards the bodies in order to retrieve them, perform burial rites, and all that," Wingo testified. "Specialist Frace attempted to stop them." Lorance approached Frace with a new order. "He told Frace not to stop them," Wingo told the judge. He said "go ahead and let them take the bodies now."

As the villagers moved in, Private Skelton stood back a few meters, watching with Lorance from under the shade of the tree. Lorance told Frace to use the radio and call Ghariban. Lieutenant Clay Comer, the fires support officer, had been repeatedly calling to see if the dead men's biometrics had been taken yet.

"Say exactly what I say," Lorance told Frace. "I've worked in the [tactical operations center] before. I know what [Ghariban] need[s] to know so that they don't ask questions."

Following the lieutenant's order to lie, Frace said over the radio, "The bodies were removed before we did a BDA."

Captain Swanson was standing directly behind Clay Comer in the operations center at Ghariban.

"At that point, I personally got on the radio," Swanson testified. I asked, "Did we enroll them in the SEEK? Did we find anything?"

Frace was in a bind. "[Swanson] asked [me] if we had put the dead persons, that was, the [Afghan] civilians, into the SEEK," Frace told the judge. Frace had just followed a direct order from his superior, Lieutenant Lorance, to lie. Now *Lorance's* superior, Captain Swanson, was asking Frace a direct question, the truthful answer to which would contradict Lorance's lie. Before Frace could respond, "Lieutenant Lorance grabbed the hand mic from me."

Lorance's response was vague, Swanson told the judge.

"You're not answering the questions that I asked," Swanson challenged Lorance. The soldiers standing next to Swanson watched what was happening. Lying to a superior officer was a court-martial-level offense.

A soldier uses a Handheld Interagency Identity Detection Equipment (HIIDE) system to scan the iris of an Afghan man, in June 2012. "The system allows soldiers in the field to quickly identify whether a person of interest is on a watch list and creates reports to support further intelligence analysis," said an army official.

| *U.S. Army, photograph by Staff Sgt. Frank Inman*

The Secure Electronic Enrollment Kit (SEEK) is used to capture fingerprints from a flex-cuffed detainee. The sensitive and classified data stored in the military's Automated Biometric Identification System (ABIS) is accessible to a select few, and understood by even fewer.

| *U.S. Department of Defense*

Convicted murderer Will West's fingerprint card at Leavenworth prison, circa 1905. The incident involving the two Will Wests triggered the end of anthropometric measurements.

| *National Archives at Kansas City, Missouri*

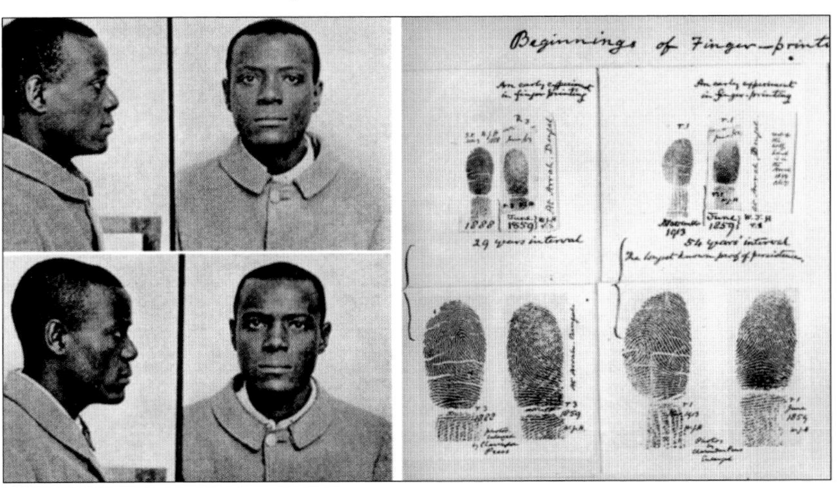

In 2004 Tom E. Bush III was appointed assistant director of the FBI's Criminal Justice Information Services (CJIS) Division, the bureau's high technology hub, where he oversaw biometric identification systems and a $1 billion annual operating budget. Bush serves on the board of the Biometric Technology Center in West Virginia.

| *Federal Bureau of Investigation*

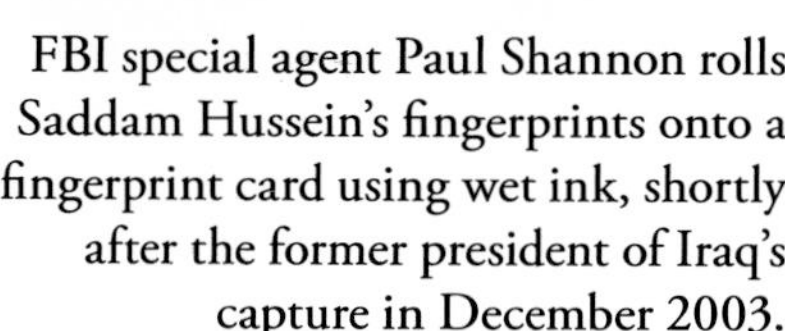

FBI special agent Paul Shannon rolls Saddam Hussein's fingerprints onto a fingerprint card using wet ink, shortly after the former president of Iraq's capture in December 2003.

| *Federal Bureau of Investigation*

Mohammed al-Qahtani, aka the Falconer, was likely meant to be the twentieth hijacker for the 9/11 terrorist attacks. His biometric data was captured by an FBI team led by Paul Shannon, at a prison in Peshawar, Pakistan, in December 2001.

| *Federal Bureau of Investigation, photograph by Paul Shannon*

Biometric Automated Toolset (BAT) systems in Fallujah, Iraq, in 2007. Interoperability problems plagued biometric capture systems from the beginning. "The original BAT was Beta when the world was VHS," lamented John Woodward, former director of the Department of Defense's Biometrics Management Office.

| *U.S. Department of Defense, photograph by Cpl. Joel Abshier*

Elementary school drawing by James Oliver Twist.

| *Collection of John Twist*

First Platoon's initial stop before combat was the Manas Transit Center in Kyrgyzstan. Here, a youthful-looking trio, including Pfc. Zachary Thomas and fellow paratroopers, build a snowman in February 2012, the last of the fun and games.

| *Collection of Zachary Thomas*

Farmer Habibullah Naim, a witness to the Kandahar Massacre and whose family members were killed, offers testimony to Afghan president Hamid Karzai in March 2012. Naim was later killed by the Department of Defense in an airstrike.

| *Collection of the Government of the Islamic Republic of Afghanistan*

First Platoon at Forward Operating Base Pan Kalay, in Maiwand District, before heading to Check Point AJ.

| *Photograph by Aaron L. Deamron*

Staff Sgt. Daniel Williams and the bomb dog on site. Sometimes, when it got too hot, the dog lost its ability to smell.

| *Collection of Daniel Williams, photograph by Christopher Murray*

A soldier collects a DNA sample on an Afghan villager in 2012. The Department of Defense's goal was to capture biometrics—including fingerprints, iris scans, facial images, and DNA—on 80 percent of Afghanistan's twenty-five million people.

| *U.S. Army, photograph by Sgt. Kimberly Trumbull*

Strong Point Payenzai (center, triangle) as seen from space. Sarenzai village is located in the upper left, Payenzai village is at the lower right.

| *National Oceanic and Atmospheric Administration*

A soldier collects a DNA sample for biometrics from a deceased insurgent. The idea behind the biometrics program was to separate friend from foe.

| *U.S. Department of Defense*

A nonplussed Pfc. Zachary Thomas, age eighteen, burns human waste at Check Point AJ in the winter of 2012. Most of First Platoon's paratroopers joined the army straight out of high school.

| *Collection of Zachary Thomas*

Pfc. Lucas Gray (left) and Pfc. James Twist (right) ready for a foot patrol at Strong Point Payenzai, outfitted in dozens of pounds of gear in oppressively hot conditions. The brainchild of Gen. David Petraeus, the idea was "Walk. Stop by, don't drive by . . . Take off your sunglasses." For soldiers, the consequences were life and limb.

| *Collection of John Twist*

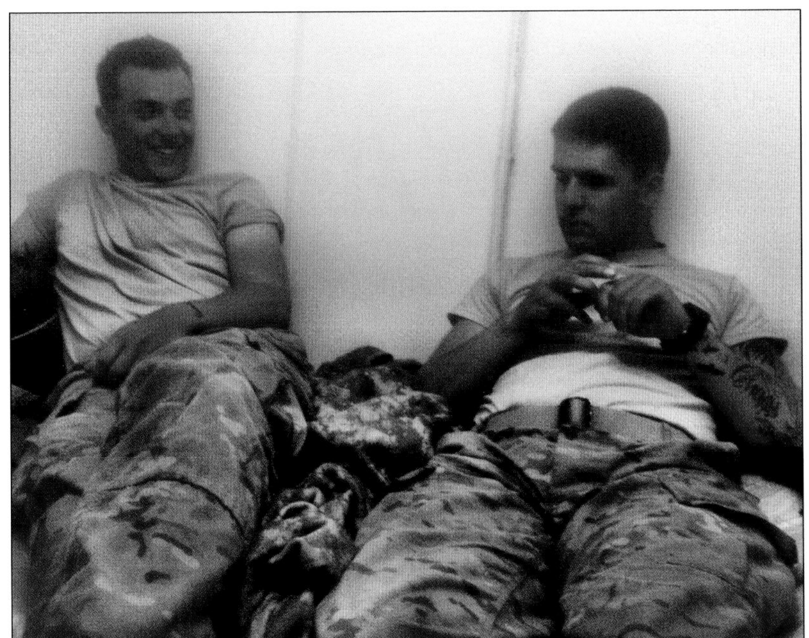

Spc. Brian Bynes (left) and Pfc. Samuel Walley (right) shortly before Walley lost two limbs to an IED buried in the ground. The word "Chaos" can be seen tattooed on Walley's arm.

| *Photograph by Alan Gladney*

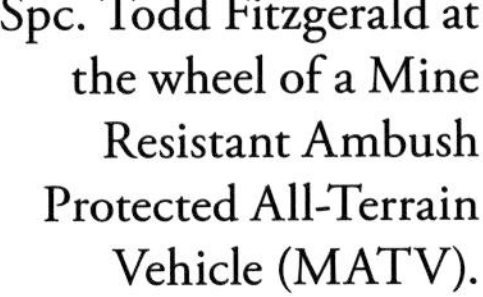

Spc. Todd Fitzgerald at the wheel of a Mine Resistant Ambush Protected All-Terrain Vehicle (MATV).

| *Photograph by Cole Rivera*

Staff Sgt. Daniel Williams, Pfc. Nathan Perkins, Spc. Anthony Reynoso, and Spc. Brian Bynes before a patrol.

| *Collection of Nathan Perkins*

A mud-brick compound between Siah Choy and Payenzai. Note the American flag flying over the Grape Hut of Woe, at center rear.

| *Collection of Kevin H.*

Abdul Ahad was assigned Biometric Identification (BID) B28JM-UUYZ by the Department of Defense and placed on a highly classified kill list. The discovery of this man's true identity in 2019 reveals as much about the military's secretive biometric programs as it does about the future of rule of law in the United States.

| *U.S. Army, National Ground Intelligence Center*

Spc. Dallas Haggard on patrol in a grape row shortly before 1st Lt. Dominic Latino and Pfc. Mark Kerner were hit by an IED near Payenzai village. The dense foliage exceeded Haggard's height, providing perfect cover for insurgents.

| *Collection of Dallas Haggard*

Tower Two inside Strong Point Payenzai on July 1, 2012, the day 1st Lt. Clint Lorance illegally ordered Spc. Matthew Rush to shoot at villagers walking to their homes. Rush later told a military judge he defied the order, saying, "'I'm not doing that.' I [could] obviously see that they're kids."

| *Photgraph by Zachary Thomas*

Growing grapes in the desert would be impossible if not for the local invention of the grape row. Here, soldiers climb up and over these tall mud-brick walls, constructed row after row, and with trenches dug down into the water table below.

| *U.S. Department of Defense, photograph by Sgt. Matt Young*

A soldier rests on top of a grape row in Zhari District, in the winter of 2012, before vegetation obscured visibility and complicated situational awareness. First Platoon was instructed to avoid IED-laden roads in favor of farmers' grape rows, but eventually these too were booby-trapped with explosives.

| *U.S. Army, photograph by Spc. Jason Nolte*

Haji Mohammed Aslam (father) and Ghamai Abdul Haq (son) were identified by the U.S. Army as the two Afghan civilians whom 1st Lt. Clint Lorance ordered killed on July 2, 2012. They are seen here minutes after the shooting, before villagers took their bodies away, photographed by helicopter pilots.

| *U.S. Army*

CHARGE III: VIOLATION OF THE UCMJ, ARTICLE 118

SPECIFICATION 1: In that First Lieutenant Clint A. Lorance, U.S. Army, did, at or near Strong Point Payenzai, Afghanistan, on or about 2 July 2012~~, with premeditation~~, murder ~~Mr. Haji Mohammed Aslam~~ a male of apparent Afghan descent (KWO 17 Apr 13) by means of shooting him with firearms. (KWO 17 Apr 13)

SPECIFICATION 2: In that First Lieutenant Clint A. Lorance, U.S. Army, did, at or near Strong Point Payenzai, Afghanistan, on or about 2 July 2012~~, with premeditation~~, murder ~~Mr. Ghamai Abdul Haq~~ a male of apparent Afghan descent (KWO 17 Apr 13) by means of shooting him with firearms. (KWO 17 Apr 13)

The army's charge sheet in its case against 1st Lt. Clint Lorance identified the two deceased Afghan villagers by name, then mysteriously crossed out each name and replaced them with "a male of apparent Afghan descent."

| *U.S. Army*

One of the men later ensnared in the double murder case against Clint Lorance was Ghamai Mohammad Nabi, assigned BID B2JK9-B3R3 by the Department of Defense after being detained "for suspected terrorist activities."

| *U.S. Army, National Ground Intelligence Center*

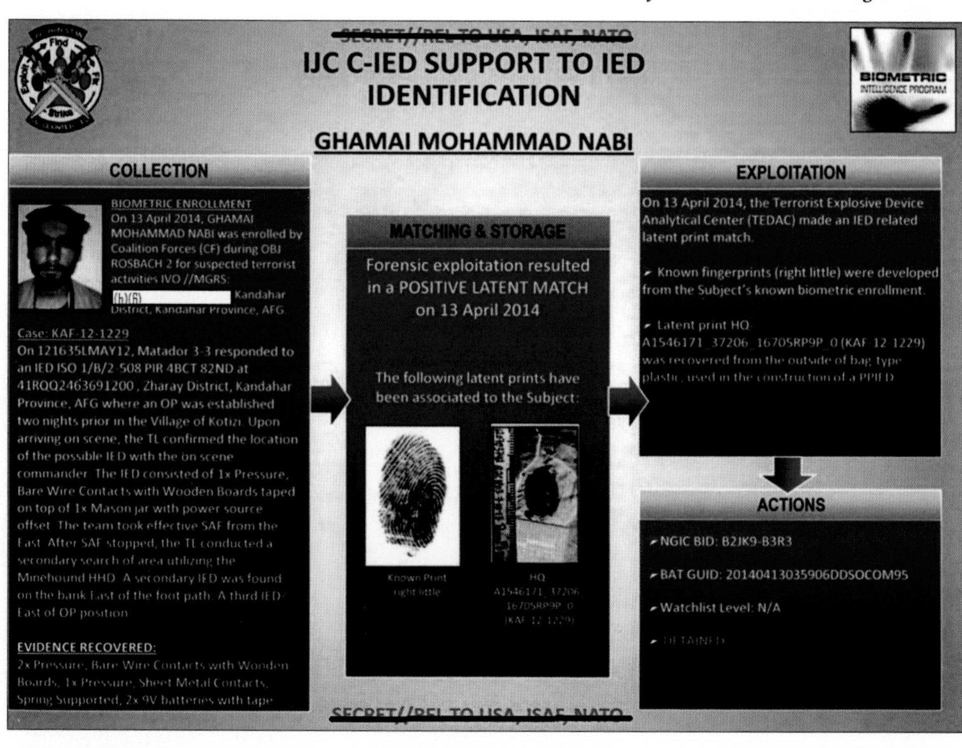

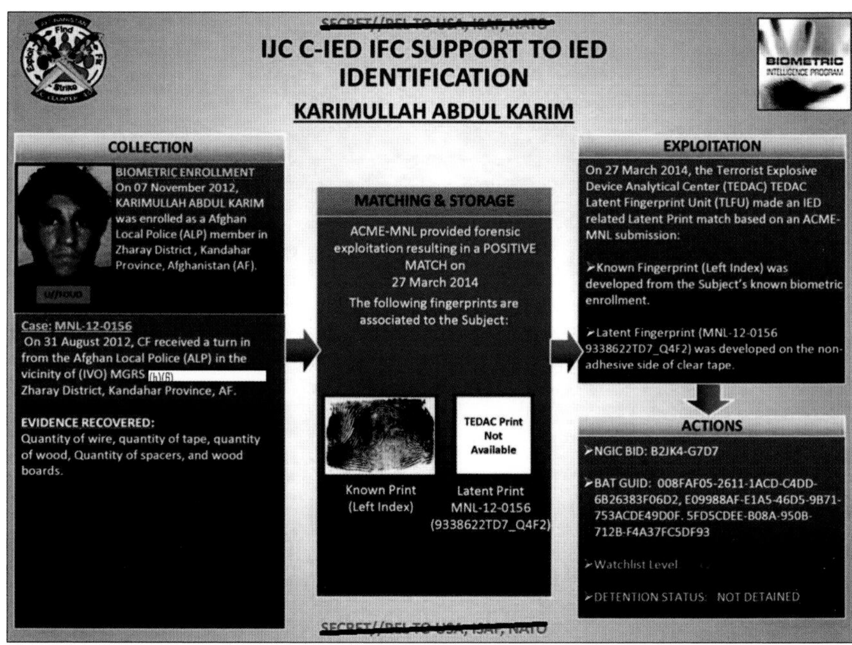

Karimullah Abdul Karim was assigned BID B2JK4-G7D7 after his fingerprints were pulled from a bomb. The FBI later concluded Karim had likely contaminated evidence as part of his official duties with the Afghan Local Police. Misinformation about this man's true identity would influence a pardon decision made by President Donald Trump.

| *U.S. Army, National Ground Intelligence Center*

Pfc. Samuel Walley, seen here with his parents, is visited by President Barack Obama as he recovers from bomb blast injuries at the Walter Reed National Military Medical Center in 2012.

| *Collection of Samuel Walley*

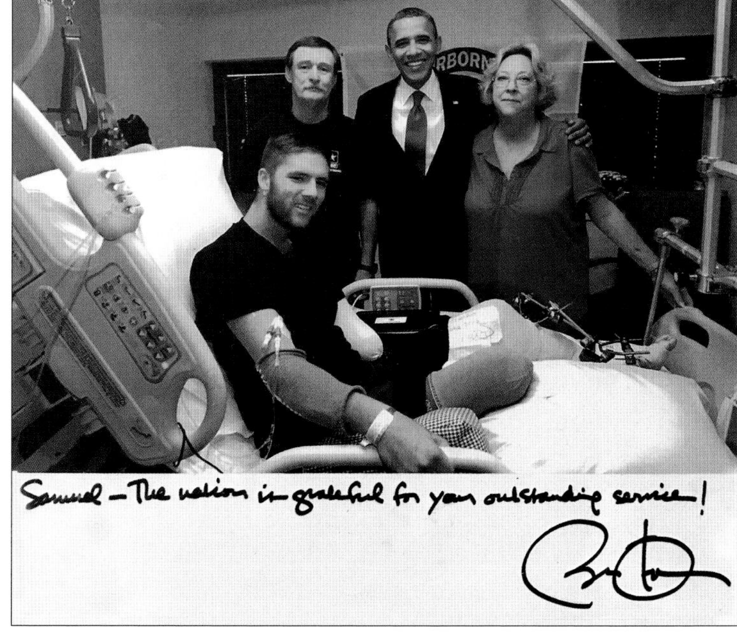

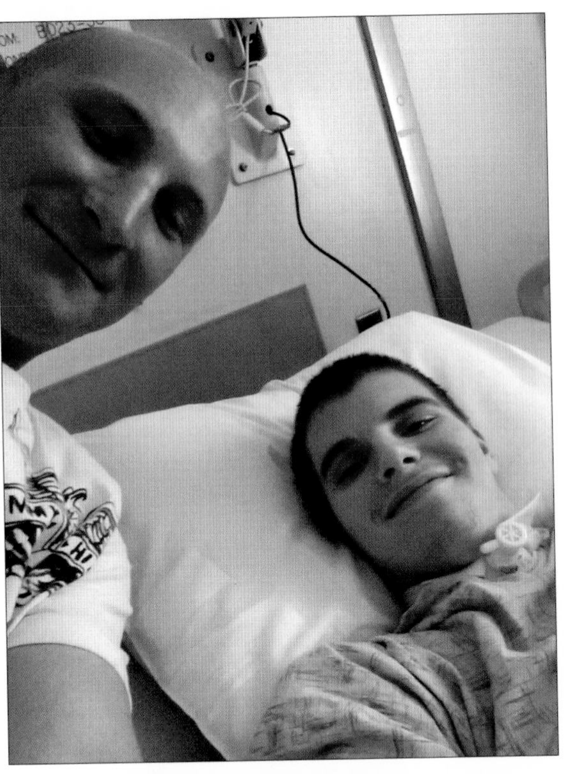

Daniel Williams visits with Matthew Hanes at a rehabilitation facility in Florida. Hanes, who served as First Platoon's minesweeper, was paralyzed by a gunshot wound to the neck in June 2012 and died of a blood clot in 2015.

| *Collection of Daniel Williams*

James Twist, age sixteen, poses in his father's army uniform from the Vietnam War. "It fit him perfectly," John Twist recalled.

| *Collection of John Twist*

Funeral of James Oliver Twist, October 2019.

Author collection

The Justice Center in Parwan was never completed. Conceived as the centerpiece for U.S. government Rule of Law Programs in Afghanistan, it only ever amounted to rusting rebar and partial concrete pours.

Special Inspector General for Afghanistan Reconstruction

Clint Lorance, granted a full pardon by President Donald Trump in November 2019, is greeted by the president on stage at a Republican Party of Florida fundraising event the following month.

| *Official White House Photo by Joyce N. Boghosian*

Inaugurated in 2018, the Biometric Technology Center in West Virginia is the first joint biometric initiative of the Department of Defense and the Department of Justice on U.S. soil. "With the opening of the BTC, the DoD and FBI will be able to work in collaboration to carry out operations and technical innovations to identify threatening or dangerous individuals," said an army official.

| *U.S. Army*

"I repeated . . . the questions," Swanson told the judge. "Did we do BDA? Did we enroll them in SEEK? Did we find anything?"

Lorance hesitated. Then he lied.

"The response that came back was 'We could not do BDA because the bodies have been taken away; they've been evacuated,'" Swanson told the judge.

Captain Swanson accepted what he had been told. The platoon had two detainees to process, Mohammad Rahim and a second man who had been detained in the village. "My next immediate concern was to have the detainees brought to Ghariban and to begin processing them immediately, to take them to Pasab."

On the way back to the strongpoint, Specialist Leon approached Private Skelton. He handed him a rolled-up piece of cloth.

"He gave me a shawl," Skelton told the judge. "He told me that there were personal effects inside the shawl." Items that Leon and Wingo had retrieved from the bodies of the dead motorcycle riders. Inside, Skelton said, "[There] was a Taskera [identity] card; some scissors; a flashlight; a vegetable or two. It's part of my job to go ahead and collect those items; take a look for any intelligence value, like the Taskera card, and to bag them if they needed to be used for evidence. For an IED identification, fingerprints, stuff like that. We can take them to a facility in Kandahar [or] at FOB Pasab, and they can do fingerprint analysis, so we have a baseline to go off of—if we find fingerprints on an IED," Skelton said at trial, explaining to the military judge how biometrics worked.

Back at Strong Point Payenzai, the platoon returned through the entry control point.

"I met up with Sergeant Ayres and Sergeant Herrmann, who came in the TOC," Williams testified. "They vented their frustration with the situation. [The soldiers appeared] distraught. Some visibly shaken. I would say Sergeant Ayres and Sergeant Herrmann were pretty angry. Sergeant Ayres looked sick."

Williams had seen much of what had happened through the Cerberus camera. He saw the riders, the killings, and the mayhem that ensued. Later, he described what he saw to the military judge.

"Soon after the shooting, the rest of our element made it into the village. I couldn't see them pull anybody out of homes or anything, but I did see them—a group of people massing by [a] large tree. I couldn't see them too well because of [a] wall that [was] there. But you could definitely tell it was a large number of people because of their heads that were moving by the tree."

Williams saw Leon and Wingo conduct the old-school BDA. "The way I recognize[d] them was through their uniforms, through their bodies, through their mannerisms, through their faces that I could see through the camera," he testified. "They moved out to the two bodies and began searching them; for anything they could [find] on them. [Doing] BDA; battle damage assessment."

Now the patrol was back. There was paperwork to fill out. A timeline to unspool.

"First Lieutenant Lorance came in [to the TOC] and said, 'Woo-hoo, that's fucking awesome.' [We NCOs] saw it and we were not amused. And [Lorance] went back in the back of the tent, dropped off his K [his kit], came back out, and said, 'OK, you guys obviously are pissed off. Give it to me—you know, just give me—tell me what we did wrong?'"

There was a pause. "[Lorance] said, 'It's OK.' That he knows how to report it up to ensure that nobody gets in trouble," Williams testified.

Two military-age males had been detained. These prisoners now needed to have their biometrics taken by Private Skelton so the automated machine, the SEEK, could determine if either of these detainees was a most wanted Taliban. If either villager was listed on the biometrics-enabled watchlist, the BEWL; the Joint Prioritized Effects List, the JPEL; or to be on the lookout, the BOLO, list.

The name of the first of the two detainees is not noted in the record. After the two motorcycle riders were killed, this man was picked up in the village by the platoon for acting suspiciously. He was not armed. The second prisoner was Mohammad Rahim. His name and his identity play a key role in the final outcome of this story.

At Strong Point Payenzai, Private Twist was ordered to keep the prisoner Mohammad Rahim under guard.

"He kept trying to talk to me," Twist recollected in 2019. "I don't speak his language, and he didn't speak mine so I said something like 'Stop talking' or 'Shut up.' But he kept talking. It took me a while to figure out he'd been shot in the arm. Weird. He didn't act like he was in pain. What kind of person doesn't feel pain?"

Private Skelton finished taking the biometrics of the first detainee. Now he began taking the biometrics on Mohammad Rahim. He later explained to the judge what processing a detainee entails.

"Upon returning to the strongpoint, we had a couple of detainees that I was processing further into the SEEK, getting their biometric data before we started transporting them up to Ghariban," Skelton testified. "While we were doing that, Lieutenant Lorance told me to not mention about the dead bodies, that we didn't do a BDA. Remembering exact words, I don't remember [but] I was told to not include that information [in my reports]."

After Skelton finished taking Mohammad Rahim's biometrics, he traveled via armed vehicle with Lorance and the two prisoners to Ghariban.

At Ghariban, Captain Swanson was walking back and forth between the COIST room and the operations center, analyzing the signals intercepts that had come in. He got on the phone with Major Ferris, in the S2 shop at Siah Choy.

"We were trying to make sense of what happened that morning," Swanson testified.

In passing he saw Lieutenant Lorance.

"I briefly asked him about the engagement. Nothing changed in his response," Swanson told the judge. "After that, at some point I ran into Sergeant Bransford, who was my COIST intelligence NCOIC [noncommissioned officer in charge]."

Sergeant Bransford told Swanson that Private Skelton had requested to meet with him.

"So after I talked to [Bransford], I felt that it was necessary to speak to Private Skelton," Swanson said.

Swanson brought Skelton into his private office. Private Skelton had bagged up the evidence that Leon and Wingo retrieved from the bodies. He pulled it out of his pocket and handed it to Captain Swanson.

"He pulled a bag out of his pocket that had a bunch of personal effects in it, like a wallet; a small knife that's used to slice poppy; maybe like keys, photos, just a small bag, plastic bag of personal effects," Swanson testified. Clearly the results of a body search, an old-school BDA. "I came out of my office convinced that I'd been lied to; that the reporting was messed-up; that the engagement might have been unlawful, and so my next stop immediately was to call Lieutenant Colonel Howard, my squadron commander. I called him. I waited by the phone and he called me back pretty quickly and he told me that Lieutenant Lorance . . . was suspended, pending an investigation."

Captain Swanson found Lieutenant Lorance filling out paperwork in the operations center.

"I told him that there were concerns about his reporting and the engagement from that morning," Swanson said, "and I told him that I was giving him a lawful order not to speak to anyone who was on the patrol with him that morning."

But it was not just Lorance who was being investigated. More than half of the soldiers in First Platoon were now being looked at for a war crimes trial. Rule of law had turned its focus on them. They were being treated as criminal suspects now.

CHAPTER SEVENTEEN

DOUBLE MURDER

At 2:41 on the Fourth of July, Captain Peter Meno, U.S. Army Criminal Investigation Command (CID), Office of the Staff Judge Advocate, Regional Command South, officially opened an investigative case on First Platoon. The basis for the investigation was succinct. First Lt. Clint Lorance "while conducting a dismounted patrol from Entry Control Point (ECP) at Strong Point (SP) Payenzai to Sarenzai Village, ordered [two paratroopers] to shoot at three male Afghans riding a motorcycle resulting in the death of two."

Within hours, eleven paratroopers from First Platoon were being held as criminal suspects for war crimes, including premeditated murder. Their fingerprints would be rolled, and their facial images captured electronically for use as mug shots. "CODIS DNA [was to be] taken at this time," except one of the special agents erred and brought only two buccal-swab DNA kits with him. Over the coming days, the soldiers' biometrics would be uploaded into the federal government's databases, to include the FBI's National Crime

Information Center (NCIC) database, its Combined DNA Index System (CODIS), and the Defense Department's Defense Clearance and Investigation Index (DCII) for past and future match-hits.

At the CID offices at Kandahar Airfield, the recent suicide of a soldier, a case already being investigated, took precedence. If you recall, the CID investigates all felony crimes and serious violations of military law for the U.S. Army, which is why it is often called the army's FBI. Three days after the shooting, Special Agent Rodolfo Rios and Special Agent Kevin Mitchell boarded a Black Hawk helicopter and flew to Forward Operating Base Pasab. At 8:20 A.M., they sat down with 1st Lieutenant Jaridd Leute, who had been assigned the job of briefing the CID agents on all that was known about the killings. Statements had been taken from every member in the platoon who had been directly or indirectly involved.

Later that morning, at 10:45 A.M., Rios and Mitchell climbed into a convoy of armored vehicles and drove approximately seven miles to Siah Choy. There, several of First Platoon's soldiers were being held as criminal suspects. Others were detained at Ghariban. In addition to premeditated murder, the crimes being investigated included making threats against civilians, changing the rules of engagement, lying to the command, shooting at unarmed civilians, and attempted murder.

At 11:30 A.M. Rios and Mitchell met with Major Jeremy Willingham, the Army regulation (AR) 15-6 investigating officer from the 82nd Airborne. Willingham gave the special agents sworn statements that had been obtained from the paratroopers by 1st Lieutenant William Charles Tinsley the afternoon of the killings.

After a briefing with Lieutenant Colonel Howard, it was decided that due to limited resources at Ghariban, the soldiers who were being detained there should be brought to Siah Choy for interviews. The investigation was approved to proceed.

The soldiers were unaware they were being investigated for war crimes.

"In the beginning I had no idea we were in trouble," Haggard remembered in 2019. "Nobody did."

"They separated us as soon as we got to Ghariban," remembered Williams. "We all gave the exact same story. The same version of events. Except Lorance. You have to remember, we barely knew him."

"The investigators were not really listening to what we said," Zettel recalled.

At Kandahar Airfield, Special Agent Sterling Brown interviewed the detainee Mohammad Rahim, the man with the gunshot wound to the arm. Rahim told investigators that he and his father were farmers. That they were working in the fields irrigating and planting crops when they were shot by soldiers on a rooftop approximately 1,000 feet away. This incident would become known as the "Second Engagement" because it occurred after the motorcycle riders had already been killed.

After the interview, Special Agent in Charge Patrick Rasmussen spent a little more than an hour "record[ing] data pertaining to Rahim, Mohammad." Private Skelton had captured biometrics on Rahim at the strongpoint. Skelton used the SEEK and a digital camera to capture Rahim's fingerprints, iris scans, facial images, and DNA. He also ran an Expray 2 test on Rahim's hands, which revealed that the prisoner's hands tested positive for homemade explosives.

With access to the Defense Department's ABIS database, and the classified data therein, Rasmussen could now look for a match-hit, to see if Rahim's biometrics linked him forensically to any previous IED events. Shortly before midnight, Special Agent Brown submitted DA Form 4254, "Request for Private Medical History," to obtain Rahim's medical records, another means by which to access a DNA match-hit. There were at least eight special agents with the army's Criminal Investigation Command working on the case. It was being treated as a major war crimes case.

•• •• ••

That same night at Siah Choy, at 10:15 P.M., a problem was discovered. Lieutenant Colonel Daniel J. Sennott, the deputy staff judge advocate for Regional Command South, Afghanistan, informed Major Willingham that the First Platoon paratroopers now being investigated for murder and other felony charges had not been advised of their legal rights. Commonly known as Miranda Rights, controlled by Article 31 of the Uniform Code of Military Justice, this legal privilege is designed to protect a person in custody from compelled self-incrimination: that they have the right to stop answering questions and the right to an attorney. The circumstances surrounding interrogation are one-sided. If an individual is going to be isolated from the outside world and subjected to professional interrogation techniques, then that person must first be made aware of their legal rights.

Colonel Sennott and Major Willingham decided that the next step in the process was to read the paratroopers their rights and to get statements from them again, or to allow each soldier to provide a "cleansing statement." It was almost midnight, and the two army officers agreed this would have to wait for the following day.

At 9:35 the next morning, the soldiers were individually read their rights. None of them refused to answer further questions, and so all of them were then given "an appropriate cleansing statement" to sign, thereby acknowledging they were "not advised of their rights as they should have, but wish to maintain their previous statement as valid." Now a new round of interrogations would begin.

"They were like hardass LAPD cops," Herrmann reflected in 2019. Special Agent Rios, a Los Angeles–area police officer, spoke with me on background about the investigation but was prohibited by the army from speaking on the record.

"They played good cop, bad cop," remembered Reynoso. "One was like, 'I'm not here to jam you up.'"

"But that was exactly what they were doing," Herrmann said.

Haggard remembers the sinking feeling that overtook him when he finally realized he was suspected of a felony crime. "They took my fingerprints. I called my dad. He worked at a chemical factory. I knew [with the time difference] he was probably coming off a shift. He answered. I said, 'Dad, remember when you and Mom got divorced and you had that lawyer? I think I need him.'"

Over the coming weeks, the CID agents interrogated the soldiers repeatedly, coming back to some of them as many as six times. Haggard was asked to write and rewrite his statement on five separate occasions. "They kept trying to get me to change what I was saying." In each of Haggard's declassified statements, he presents the same set of facts.

"The reason I said the same thing [about what happened] . . . why all of us said the same thing [about what happened], is because it was what happened," Haggard told me in 2019.

In addition to pursuing felony charges against Lieutenant Lorance, the agents also began building criminal cases against eight of First Platoon's combat infantry soldiers and NCOs: Private Skelton, who initially fired at the men, but missed; Private Shilo, the gunner in the gun truck; Private Carson, Specialist Haggard, and Sergeant Herrmann, the three soldiers involved in the "Second Engagement"; Specialist Zettel and Specialist Rush, the two marksmen involved in the harassing-fire incident from Tower Two on July 1; and Sergeant Ayres, who neglected to report Lieutenant Lorance for Uniform Code of Military Justice violations, including an order to another soldier to lie on his behalf.

"There are a lot of moving parts in this convoluted [report]," wrote Special Agent in Charge Patrick Rasmussen in a law enforcement sensitive report.

"Excellent job managing such a huge case," wrote the team chief, Special Agent Brian Marshall.

On July 7, 2012, five days after the killings they were investigating, Special Agent Rios and Special Agent Mitchell visited the crime scenes at and near Strong Point Payenzai for the first time. After arriving by armored vehicle, they went to the makeshift tactical operations center. Since there was no electricity in Payenzai, the TOC was a hand-hewn plywood desk covered in camouflage netting. It was there that the investigation took an unusual turn, one that has never been reported before.

According to the declassified CID Summary of Investigative Activity Log, "Upon exiting the TOC, Special Agent Rios and Special Agent Mitchell are approcahed [*sic*] by Private Skelton."

What was Private Skelton doing back in the battle space at Payenzai? Everyone else from First Platoon who was facing murder, attempted murder, and other felony charges was being detained at Siah Choy. They had been separated from other paratroopers, and had been ordered not to speak to anyone from their platoon. And yet here was Private Skelton, back at Strong Point Payenzai, doing his job as COIST.

"Special Agent Mitchell asked Private First Class Skelton if he was still aware of his Article 32B Rights." Skelton said yes. The rest of the details surrounding this anomaly in the investigative process remain classified.

According to the record, Private Skelton then "escorted" Rios and Mitchell to the crime scenes. This action was so unusual that when Special Agent in Charge Patrick Rasmussen was reviewing the investigation, he flagged this event for explanation. The explanation of this event also remains classified.

At ten o'clock that morning, Rios and Mitchell walked to the scene of the killings. The First Engagement, in CID parlance. With them was Afghan translator Mirwais Milyar. They found a stain in the dirt, eight inches in diameter. They photographed the site, the tree, the mud-brick walls, and the nearby grape rows. They noted the "line of sight" positions of the individual platoon members involved in the crime. They measured the distance from Strong Point Payenzai to Sarenzai Village as 1,000 feet. "Security concerns imposed time restrictions on the agents to complete a thorough examination."

Next, they walked 350 feet down the dirt road, heading west. A twelve-foot-wide road took them through the mud-brick homes of Sarenzai Village. They went past a five-foot-high, "brown in color wall . . . made of the same mud like structure" as the buildings. Past the spot where Hanes had been shot in the neck two weeks before. They located a "not well kept . . . vacant or unoccupied" mud-brick structure, from which the shots in the Second Engagement had been fired. They climbed up on the roof and photographed the surrounding area with a digital camera.

Then came another pivotal moment in this complex and highly charged criminal investigation.

At 10:53 A.M., the special agents were approached by a villager. Through translator Milyar, the agents learned this man was the son and the brother of the two motorcycle riders killed in the First Engagement. His father was the village elder, he said, and was named Haji Mohammed Aslam. His brother was named Ghamai Abdul Haq. The third rider, the man dressed in white who ran away, was his uncle, this villager said. That his uncle was named Haji Karimullah, and lived in the nearby village of Pashmul. As proof of identity, this villager showed Rios and Mitchell his Afghan-issued identification card, Taskera #1571683. The investigators recorded the name of this villager in their logs.

His name was Abdul Ahad.

Over time, the significance of the true identity of this man—

and what this "truth" says about the strength, and frailty, of the rule of law—would astound.

At Siah Choy, Captain Michael Hanna was concerned. Lieutenant Lorance "seemed to be unaware of the situation he was in," Hanna told investigators, as if Lorance didn't realize he was facing a felony double murder charge. Lieutenant Brian Delgado, accompanying Lorance after his weapon was taken from him, described him as "emotionally detached." That he "appeared to be shell shocked, like if the incident never happened."

Second Lt. Katrina Lucas told investigators, "After [Lorance] returned from SP Payenzai [he] was different. He did not seem like he was embarrased [*sic*] about the event, or for being relieved of his position. He was very nonchalant and did not care about anything."

Second Lieutenant Lucas Pierce was assigned to be Lorance's escort, or guard, at Siah Choy. "He would occasionally make inappropriate jokes," Pierce told investigators, "comments about killing people, said in a joking and lighthearted manner. I never saw any kind of remorse, sadness, mourning, or solemn behavior that would be anticipated after an event like a CIVCAS. He instead was upbeat, joking and always smiling." Lorance's reaction to having ordered the deaths of two civilians was so aberrant, Pierce said, the other officers referred to him as "LT Crazy Pants."

As days passed, Captain Hanna's concerns about Clint Lorance's behavior worsened. "At one point, Clint made the comment, 'Man, when I get back I am going to kill every motorcycle I see,'" Hanna told investigators. "He was laughing. This laughter was not the typical I am having a wonderful time at a July 4th cookout with my family type laughter but rather a more evil manic laughter which alarmed me."

Captain Hanna expressed fear for his own safety. "I am nervous around him," he told CID. "[I] don't want to be the guy who tells him any bad news and end[s] up strangled by him because of it. He has no remorse for anything he did . . . I realize they took his weapon but he still has access to other weapons."

On July 9, one week after the killings, Special Agent Mitchell advised 1st Lt. Clint Lorance of his legal rights. Lorance asked for a lawyer. Later that day, the army's trial lawyer sat down with a team of six special agents from CID to figure out how best to build their case, and "what subjects to list for what offense(s)." After the session was over, they agreed to build criminal cases against nine paratroopers from First Platoon.

Special Agent in Charge Rasmussen sent out a memo to his team of investigators. "Excellent brainstorming session," he wrote, "everyone is tracking on the progress of the investigation."

On July 14, twelve days after the killings, CID investigators noted for the first time that Haji Karimullah should be interviewed. "Haji Karimullah is an important interview, as he was uninjured as a rider on the motorcycle while his relatives were killed. Ensure he is on the [interview list]," Rasmussen noted in a quality-control review. But as had been the case after the Bales killings, local government had very little knowledge of who its citizens were, where its people lived, or how to get in touch with them during a felony murder investigation by the Americans.

At Pasab, Lieutenant Colonel Frank Harrar began working with the Afghan Local Police to locate and interview Haji Karimullah. Four days later, Harrar followed up with the local police. They were still waiting on a telephone number, he learned. On July 22, Afghan interpreter Mirwais Milyar made contact with Haji Karimullah, who agreed to meet with army investigators the following day at ten A.M. The location chosen was the Pasab District Center.

At 10:55 A.M. on July 23, three full weeks after the murders,

Special Agent Mitchell interviewed Haji Karimullah, through interpreter Milyar. The testimony Haji Karimullah gave investigators was reported by the agents, as follows:

> Mr Karimullah related he was a guest of [his brother] Mr [Haji Mohammed] Aslam and [his brother's son] Mr [Ghamai Abdul] Haq at their residence in Panjawa'i. They were traveling from their residence to their grape fields in Sarenzai when they saw a military truck parked in the road. Mr Karimullah related due to the road conditions they could not ride fast, and were not riding fast when they saw the truck. Mr Karimullah said almost everyone in the local villages owned a motorcycle because they were cheap, and able to ride on the roads easily. He also said the route they had taken from Panjawa'i to Sarenzai was the only route available and that everyone used it. It was very common to ride a motorcycle on that road.
>
> They next heard warning shots and stopped the motorcycle. Mr Karimullah related they did not see the patrol in the grape fields until after they stopped the motorcycle. Mr Karimullah and Mr [Haji Mohammed] Aslam [his brother] dismounted the motorcycle while Mr Haq [his nephew] stayed on.
>
> Mr Karimullah yelled to the ANA that he was good and wanted to tend to his grape fields and house, but was too far away for them to hear. He decided to approach the ANA with his hands up to show he was unarmed, and ask where they wanted him to go to talk, but was met with a hand gesture to return to the motorcycle.
>
> Mr Karimullah and Mr Aslam conversed and expected to be approached by either ANA or U.S. Forces for an interrogation and an opportunity to identify themselves as locals and not Taliban. Mr Karimullah stated approximately 3–5

> minutes after they stopped the motorcycle that shots were fired from the truck.
>
> One shot almost struck Mr Karimullah, and another struck Mr Haq [his nephew] in the neck. Mr Karimullah and [his brother] Mr [Haji Mohammed] Aslam ran to Haq and attempted to stop the bleeding. Mr Karimullah told [his brother] Mr Aslam they needed to get help, but Mr Aslam refused to leave his son's side.
>
> Mr Karimullah then ran into the village to find help. Mr Karimullah found family and they ran toward the motorcycle where they were stopped by the U.S. Patrol.
>
> The U.S. Patrol told them they had to wait because they wanted to see if the riders were criminals. At that time, Mr Karimullah was unaware [his brother] Mr [Haji Mohammed] Aslam had also been shot. He stated he was running and yelling and did not hear any more shots, but was told by family that more were fired. Once stopped by the U.S. Patrol, they were not allowed to recover the bodies for approximately 45 minutes.

Karimullah's story lined up precisely with testimony from Specialist Fitzgerald, Private Thomas, Private Skelton, Sergeant Williams, and others. Testimony that was taken just hours after the crime. Haji Karimullah was "adamant" with special agents that he, his brother, and his nephew were not members of the Taliban. They were grape farmers, Haji Karimullah said.

> Mr Karimullah related he had been to the ECP of SP Payenzai approximately 5 or 6 times prior to this incident and had dealt with the U.S. Paratroopers there, asking if he could irrigate his grape fields. He said he never had any problems with the U.S. Paratroopers prior to this incident, but that his

> family is now scared of the U.S. Paratroopers and has sought medical treatment to deal with the mental problems acquired because of this incident. [The deceased] Mr [Haji Mohammed] Aslam's two other children (31 & 34 yrs old) are now tending to the grape fields since their Father was killed.

After recording Haji Karimullah's testimony, Special Agent Mitchell explained that money could be paid to him for his loss. This kind of solatia payment was commonplace, Mitchell said. The family members and victims of Staff Sergeant Bales's murder spree—approximately six miles south, four months before—had received $970,000 in solatia payments as part of the Defense Department's consequence-management efforts.

Haji Karimullah told the special agents he was not interested in taking their money. "Mr. Karimullah stated his family had not accepted the condolence payment as they did not want to forgive the U.S. for the deaths of their family members."

The investigation dragged on. For those not ensnared in the investigation of the killings, morale at Strong Point Payenzai felt interminably low.

"Being investigated for war crimes?" Private Twist questioned the very idea. "Not possible." Except it was. The army had turned on First Platoon. That this was happening felt as ruinous and arbitrary as stepping on an IED, Twist said.

"It wasn't just Payenzai that was cursed," said Private Thomas. "It felt like our platoon was cursed."

"The one school we built next to Ghariban ended up getting destroyed by us," Twist lamented. "We hit it with our own missile. We weren't supposed to say that's what happened, but it did. The

Taliban took it over and were using it to plan IED attacks. [It] sums up our deployment. Everything good we did ended up getting destroyed."

The criminal investigation lasted six months. In the end, investigators found no probable cause to charge any of the privates, specialists, or noncommissioned officers with murder or attempted murder. Four received disciplinary charges.

On January 8, 2013, trial counsel Captain Kirk Otto gave his legal opinion as follows: Sergeant 1st Class Ayres was to be charged with Dereliction of Duty for failing to report that Lieutenant Lorance ordered Sergeant Williams to lie about shooting harassing fire into Payenzai Village. Staff Sergeant Herrmann was to be charged with Failure to Obey a General Order for firing the Carl Gustaf from the strongpoint in the direction of a Taliban shooter because this violated the rule of proportional force. Specialist Rush was charged with Discharge of a Firearm in Circumstances Such as to Endanger Human Life for shooting harassing fire into Payenzai Village; investigators concluded he should have refused to follow those orders. Specialist Zettel was charged with Willful Discharge of a Firearm in Circumstances Such as to Endanger Human Life even though he never fired his weapon during the harassing-fire incident. As for Specialist Haggard, Private Carson, Private Shilo, and Private Skelton, Captain Otto believed criminal investigators had found no probable cause for any offenses.

First Lt. Clint Lorance was to be charged as follows: "two incidents of murder, one incident of attempted murder, one false official statement, one obstruction of justice, and one willful discharge of a firearm in circumstances such as to endanger human life."

In the end, the identity of the murder victims mattered little to the Defense Department. On January 1, 2013, the case was submitted for final review. The special agent in charge flagged as unacceptable the fact that the murder victims were still being called "Local National Victims."

"Final can't be submitted until the mandatory fields are filled in," the agent wrote. The following week, names were uploaded into the system. On January 15, these names appeared on the original charge sheet as:

— premeditation, murder [of] Haji Mohammed Aslam
— premeditation, murder [of] Ghamai Abdul Haq
— attempt to murder with premeditation, Mr. Haji Karimullah

A few days later, one last flaw was discovered. There were no birth dates on the victims, also a mandatory field in the report.

"SAC [special agent in charge] further stated based on the victims being local nationals, to give all victims the same date of birth as there is no way to know the date of birth of the individuals." As a solution, "SAC added the DOB [date of birth] of 1 Jan 1985 for all victims."

And so, in official U.S. Army records, the father and son murder victims were listed as being the same age, twenty-seven and a half years old, when they were killed.

What happened at Strong Point Payenzai is a complex, and convoluted, tale of ruin and loss. Of hypocrisy and impunity masquerading as the rule of law. For Pfc. Samuel Walley, a double amputee, and Spc. Matthew Hanes, paralyzed from the neck down, life would never be the same.

"Lieutenant Latino was in the hospital with me," Walley remembered in 2019. "He looked like Two-Face from Batman. I wanted to go see Hanes [on another floor], so he came with me when I went to see Hanes for the first time. I came rolling in, in my electric wheelchair. Said, 'Hey, Hanes, you son of a bitch.' He smiled a little. He moved a little bit. Then I realized that he was paralyzed. All he could do was move his eyes."

President Obama visited Private Walley at the hospital to honor him as a war hero. Walley mustered a smile.

"President Obama talked to me for the longest time," he remembered. "He wasn't in a rush. He came back later, a second time. He said, 'Hey, Walley, looks like you gained some weight.'"

But Walley was in terrible shape, mentally and physically. His skin grafts were infected and the doctors told him his left leg would likely have to be amputated, which would make him a triple amputee.

"Afghanistan is one big giant sea of bacteria," Walley said. Bacteria first enters the open wounds on the battlefield, then continues to grow inside the body. "Many of us amputees struggle with MRSA [methicillin-resistant Staphylococcus aureus infections] for years."

Hanes was moved to a longer-term care facility in Tampa, Florida, where many of the paratroopers visited him.

"The thing was," said Twist, "Hanes was still the most positive person in the platoon."

"Even when everyone else around him was falling apart," Walley agreed.

In the hospital, Hanes read stories on the Internet about Chinese doctors who claimed to cure paralysis with stem cells. He began writing Facebook posts about taking a trip to Beijing. When Sergeant Williams visited him at the rehabilitation center in Florida, Hanes shared with him how excited he was to one day walk again. His mother was helping him raise money for the surgery and the trips, he said.

"When I left, I gave Hanes my good-luck red poker chip," said Williams. It had been in his pocket the duration of the deployment in Afghanistan.

•• •• ••

Over the New Year, in 2013, nine of First Platoon's paratroopers received notices in the mail. Grants of immunity and an order to testify against 1st Lt. Clint Lorance in an upcoming double-murder court-martial trial. The trial would take place at Fort Bragg in the summer of 2013.

The 986-page trial transcript exists as a public record of fact. The testimony of more than a dozen witnesses paints a portrait of Clint Lorance as an overzealous lieutenant who acted with impunity and in defiance of U.S. Army rules of engagement and general principles guiding the rule of law. In his three days in a combat zone, Lorance tormented local Afghans; issued death threats; ordered soldiers under his command to shoot harassing fire at old men, women, and children; ordered soldiers to kill civilians whom he mistook for enemy combatants; ordered a helicopter pilot to kill a civilian; lied to command to cover up crimes; and ordered other soldiers to lie to cover up crimes. As an officer inside a tactical operations center, Lorance appears to have done his job honorably. On the battlefield, it was as if something in him snapped.

In August 2013, a ten-member jury of military officers convicted 1st Lt. Clint Lorance on "two counts of murder and other charges related to a pattern of threatening and intimidating actions toward Afghans, as commander of an infantry platoon." John Ramsey in the *Fayetteville Observer* wrote, "Lorance was convicted of committing at least one crime every day he was with his unit at a small outpost in the Kandahar province of Afghanistan last summer."

It is hard to imagine that life in America could feel more treacherous than life had felt in Zhari, Afghanistan, but many from First Platoon say that was the case.

"The military community pretty much turned against us for testifying against a lieutenant," said Twist.

Things got worse. They were no longer a platoon. They were no longer a brotherhood, a tribe.

They were ostracized veterans of a war many Americans did not seem to care about, or to understand. They were individuals, alone.

"No one wanted to be associated with us," said Twist.

"Pretty much all of us got run out of the army, or quit," said Herrmann.

"They treated us like pariahs," said Williams. "Like we had the plague."

PART V

CHAPTER EIGHTEEN

THE CLUE TRAIL

It was the spring of 2014 and a former Department of Justice litigator and army officer named John N. Maher was standing in his Chicago apartment ironing a shirt. Maher was getting ready to act as a pallbearer in his cousin's funeral. The cousin, age thirty, had committed suicide.

"It was a grim day. Cold, dreary, depressing," Maher told me.

Maher's cell phone rang. He almost didn't answer it, but then saw it was an old friend calling, a fellow lawyer and former U.S. Army JAG officer named John D. Carr. So Maher picked up the phone and said hello.

"A lieutenant in the 82nd Airborne is in prison," Maher remembered Carr telling him. "He asked me to help."

After the funeral, Maher called Carr back and they had a discussion about the case. In no time, John Maher was serving as lead defense counsel for Clint Lorance in a quest to get him out of Leavenworth.

As an attorney, John Maher, who is also a lieutenant colonel in the U.S. Army Reserve, had rare expertise in an esoteric area that was about to become central to Clint Lorance's new defense: combat biometrics.

"The first thing we did was look at the charge sheet. The criminal indictment," Maher said, explaining the timeline to me. "The names of the Afghan victims were crossed off, in penmanship, and instead language was inserted: 'a male of [apparent] Afghan descent.'"

Why would the U.S. Army judge advocate general do such a thing?

"My suspicion was this had to do with biometrics," Maher said. "With the true identities of the men killed."

He shared this document with me, and it was puzzling. There, on the army's original charge sheet, were the crimes Lorance was originally accused of committing, followed by the names of the three victims. As we know, it read:

—premeditation, murder [of] Haji Mohammed Aslam
—premeditation, murder [of] Ghamai Abdul Haq
—attempt to murder with premeditation, Mr. Haji Karimullah

But the original charge sheet had indeed been altered. Just as John Maher stated, each of the victims' names had been crossed out with a line that appears to have been drawn by a human hand.

Upon seeing this, John Maher's legal training came to bear. "It was the old 'spot the issue' thing from law school," he told me. "That's an issue right there. Shouldn't we know who you're accused of killing and attempting to kill, or murdering and attempting to murder? The fact [is] the prosecutor lined that out. This told me two things," Maher said. "Number one, they knew who these people were and didn't want to necessarily get into it. Or, number two, they had a problem with the fact that they couldn't prove that Lorance killed these people."

Maher reviewed all the documents available to the defense. He found another clue. The army wanted to exhume the bodies. "Now, why would they want to do that? Our suspicion was this, too, had to do with biometrics." With the true identities of the men killed.

John Maher's select legal expertise was even more specific to this case. Starting in the fall of 2012—shortly after First Platoon left Afghanistan—John Maher worked in Afghanistan as one of the lead attorneys for the U.S. government's Rule of Law Program at the Justice Center in Parwan, the JCIP. This facility, if you recall, is the American-built justice-center complex designed to sit adjacent to the American-built Detention Facility in Parwan, Afghanistan's Guantánamo Bay. In the rule-of-law paradigm of law enforcement, courts, and corrections, the Justice Center in Parwan was the courts component. Starting in 2010, it was the only facility in all of Afghanistan where members of the Afghan judiciary could learn about biometrics, forensics, and Western-style rule of law. It was the crown jewel in the Defense Department's multibillion-dollar Rule of Law in Afghanistan programs.

As a principal program manager at the justice center, John Maher oversaw a team of approximately one hundred U.S. and Afghan personnel there. These were American attorneys, paralegals, and criminal investigators who trained, mentored, and advised Afghan attorneys, judges, and criminal investigators. The idea was, the Afghan professionals were to become the backbone of Afghanistan's criminal-justice system after the Americans left. Everybody at the Justice Center in Parwan had the same goal, Maher said: "prosecute Taliban bomb makers and put them in jail."

"Biometrics was the centerpiece of everything we did," Maher told me. "It was an hourly event. In those fourteen- or fifteen-hour days [we worked], we were dealing with biometrics no fewer than fifty or sixty times in an hour. And I mean that, because [we] had a number of prosecutors, with a number of cases, with simultaneous trials ongoing. I was in charge of all of that."

In 2012 and 2013, the justice center complex had still not been constructed, so the employees all worked in temporary facilities, inside Conex boxes and relocatable buildings at nearby Camp Sabalu-Harrison. Every day the Americans and the Afghans ate lunch together: lamb, oranges, tea. On occasion, an incoming mortar meant they had to take cover in a bomb shelter.

"All the men being tried at the justice center were Taliban insurgents," Maher made clear. "They shuffled them over from the detainment facility, in leg-irons. If they weren't wearing a hood, they'd spit at us. It was nasty business." Some threw shoes. "They were all battlefield pickups," Maher recalled. "People who are detained with HME on their hands, homemade explosive residue, or people who had not been killed in a gunfight" but were captured.

People like the number two most wanted Taliban—the man whom First Platoon stopped and then detained by the water pump in Sarenzai Village on June 4, 2012. On Private Walley's twentieth birthday.

"We'd put them in the Detention Facility in Parwan," Maher continued, "the idea there is not only to take them off the battlefield so they'll stop being a danger to the coalition as well as the civilian population, but also to secure intelligence information from them to help us plan and understand other operations." To build cases against other IED cell members through Identity Intelligence, or I2.

From his living room in Chicago, in 2014, Maher began building his case to get Clint Lorance out of prison. He searched for, and found, additional clues. "Like, why did Mohammad Rahim, the battlefield capture, have HME on his hands?" Mohammad Rahim, if you recall, is the detainee who was shot and captured in the Second Engagement on July 2, 2012. But the biggest clue, said

Maher, involved the government witness named Abdul Ahad. Maher found this witness to be very suspicious. "Why did [the army's] CID [investigators] take him at his word? Where are his biometrics?"

This is what Maher focused on in the spring and summer of 2014.

Maher developed a theory, he told me. The army crossed out the names of the Afghan victims because the victims "were actually bad guys." Meaning, he said, "the army wanted to hide who these individuals actually were." But how to learn the true identities of these men?

Maher had rare expertise in the U.S. government's use of biometrics in Afghanistan. He knew that databases that contained classified information, like the Defense Department's ABIS, were accessible to only a select few—and understood by even fewer. His hunch was that the army was banking on the fact that no one would ever question the identities of the motorcycle riders, let alone be able to find out who they really were. Except, John Maher believed, he knew someone who could.

Maher had a colleague who had knowledge of, and access to, the government's ABIS database. Standing in his Chicago apartment in the fall of 2014, John Maher picked up the telephone and called his colleague, a criminal investigator and Defense Department contractor with whom he'd worked at the Justice Center in Parwan. The man's name is William Carney.

"Bill Carney is *the* expert for the last ten years on biometrics in Afghanistan," Maher stated in a Starz documentary made about Clint Lorance's imprisonment, called *Leavenworth*. "He's a retired New York City police officer with whom I served in Afghanistan at the Justice Center in Parwan, and we prosecuted the Taliban insurgent fighters [together] in a Top Secret national-security court, using biometric information."

But getting Bill Carney to help out was going to be tricky, Maher told colleagues.

"Bill Carney was the man," said Don Brown, a former U.S. Navy JAG officer, former special assistant U.S. attorney, and soon-to-be member of Lorance's new defense team. "[Carney] still worked in Afghanistan on the U.S. government's national biometric enrollment program."

"I sent Bill Carney an email asking for a call," Maher recalled.

John Maher and Bill Carney had a well-established relationship. For approximately ten months in 2013, the two men worked side by side in Parwan, Afghanistan. As an investigative justice adviser, Carney taught Afghan policemen and lawyers how to use forensic and biometric evidence found at crime scenes. He trained officials with the Afghan Ministry of the Interior how to collect and use this evidence as part of the Afghan 1000 program, the indigenous element of the U.S. effort to collect biometrics on 80 percent of Afghanistan's 25 million people.

"At JCIP [I] worked as the lead investigator [and] adviser on all 'high visible' cases where biometric associations, fingerprints, and DNA are involved," Carney confirmed with me in an interview from Helmand Province, Afghanistan. But his real expertise, he said, lay in building criminal cases against networks of IED cells, which was similar to building cases against Mafia cells. "I was on the first Criminal Network Cell team in Afghanistan," Carney told me, "credited with the first successful prosecution of a ten-man team of Taliban bomb makers," who were all tried and convicted in August 2011. "These were ten individual captures," Carney said. "Not guys captured as a group. We built their associations as a cell of bomb makers from their biometrics." Because Carney worked with the National Directorate of Security, Afghanistan's CIA and FBI in one entity, a lot of what he does is classified, he said.

In 2014, when John Maher reached out to Bill Carney via email, there was no guarantee the former detective turned biometrics expert could, or would, help.

"John [Maher] hoped and prayed that based upon their previous relationship, Carney would be willing to help out," co-counsel Don Brown recalled.

"No one was more important to the top-secret prosecution program [of Taliban bomb makers] than Bill Carney," Maher told the new defense team.

"John sat [down] in front of his computer and sent an email to Afghanistan," remembered Don Brown. "It read: 'Bill, please call me.'"

John Maher believed his ability to get Clint Lorance out of prison depended on Bill Carney's assistance, he told the *Leavenworth* documentary filmmakers. "Bill Carney's involvement in the biometric development of the evidence in this case [would make] the case. The appeal [to the army courts] wouldn't be nearly as strong, or nearly as convincing, or nearly as compelling without the actual evidence that the military had this [biometric] information at its fingertips."

If the army's Criminal Investigation Command had withheld biometric evidence indicating that the men on the motorcycle were Taliban bomb makers, as Maher suspected, that would be a violation of Clint Lorance's Fifth Amendment rights. His right to due process of law. But this biometric information was highly classified and strictly controlled.

Carney could access it, Maher told me. All he had to do was "punch those names into the appropriate database."

After Maher sent the email, the suspense was palpable. "Several hours passed," co-counsel Don Brown wrote. "The hours felt like days."

Then finally, Bill Carney agreed to speak to John Maher on the phone.

"We spoke the next day," said Maher. "He was in Afghanistan and I was at my home office one mile west of Wrigley Field in Chicago. I explained the case, and that we had names of Afghan victims crossed out and in the CID agent activity summaries."

When the two men were at the Justice Center in Parwan, Maher perceived Carney as the glue that helped keep many key elements together, he told me. Carney supervised lawyers' requests for biometric information, tracked individual Taliban bomb-maker case files, and made sure the information got to where it needed to go. With the authority to apply U.S. government security markings to documents, Carney was also one of only a few people at the justice center who were what is called "burn rights" certified. This meant Carney was cleared by the foreign disclosure representative to burn sensitive information—including documents, photos, and videos—from the ABIS database onto compact discs, for Afghan court purposes.

Maher explained what happened in this critical telephone call.

"I gave Carney the names of the three motorcycle riders and the name of the witness, Abdul Ahad," Maher told me. "All we had in the beginning were their names: Ghamai, Karimullah, Ahad, and Rahim. I asked if he would punch those names into the [ABIS] database." Maher was looking for very specific forensic and biometric information. "[Had] they left their fingerprints and/or DNA, meaning their skin, [because] when a terrorist twists wires, they leave their skin on the wires or the bomb components," he specified. "[Had] these people left their fingerprints and DNA on bombs at grid coordinates where IED events were recorded?" The military grid coordinates were important, Maher made clear, because this data allowed prosecutors at the justice center to link Taliban bomb makers to specific deaths of U.S. servicemen and their coalition partners. Maher asked Carney to look into all of this.

What John Maher was telling me was remarkable. To "punch" those names into the ABIS database is not something just anyone could do. ABIS contains data that is highly classified. That Carney had unfettered access to it, and was able to share that information with John Maher without violating the law, including potentially the Espionage Act of 1917, that was nothing short of extraordinary. The two men were colleagues but also friends. Maher called Carney "Coach."

"I said, 'Coach, will you try and pull the biometric enrollment records of these named individuals? See if there are any hits?'"

Carney said he would see what he could do.

"We knew if the biometric data tied any of these four Afghans to bomb making, to IEDs that killed American soldiers, we knew that would be a game changer for Clint Lorance," John Maher said.

The day after Thanksgiving 2014, a DHL package arrived at John Maher's home. The contents, he told me, "were a smoking gun."

Bill Carney had found exactly what John Maher was looking for. And it was a stunning reveal. Abdul Ahad, the main witness whom the army had built its double-murder case against Clint Lorance around, was a most wanted terrorist, Carney said. To prove it, Carney had located the alphanumeric biometric identification assigned to Abdul Ahad, which was B28JM-UUYZ. The Abdul Ahad written about in Chapters Eleven and Thirteen of this book.

Abdul Ahad was the leader of a Taliban bomb-making cell, a man marked for death by his placement on the Defense Department's classified Joint Prioritized Effects List. He was an IED manufacturer and Taliban commander who lived near, and operated in, Zhari District. Wore a green dishdasha. Drove a red motorcycle. He had been connected by intelligence analysts to at least fifteen Taliban commanders, six of whom were also marked for death.

At Maher's behest, Carney got to work linking together members of Abdul Ahad's cell. He linked Mohammad Rahim, the detainee shot in the arm in the Second Engagement, to Abdul Ahad, he told Maher. Then came the smoking gun: Carney was able to show that Ghamai Abdul Haq, one of the motorcycle riders Clint Lorance was in prison for having killed, was also a known Taliban bomb maker. And so was Haji Karimullah, the man in all white who escaped, Carney said. Each man had left fingerprints and/or DNA on IED components that killed American soldiers, Carney told Maher. This was revelatory, Maher told me, because fingerprints don't lie.

Carney shared with the *Leavenworth* documentarians how he had performed the link analysis by searching the government's databases. "The information I had [delivered] numerous hits," he said. "In these databases were all these IED events, [and these] individuals, they all came together in networks . . . These individuals were linked to IED events within the Zhari and Panjawai Districts."

When Bill Carney returned from Afghanistan, John Maher flew to his house. It was Easter time, he recalled.

"He sat me down and presented it all to me like I was the jury," Maher told me. PowerPoints, charts, links, military grid coordinates. "It was all right there," Maher could see. "A first lieutenant, in prison for killing the enemy."

Bill Carney was able to deliver information that read like a plot twist straight out of a detective novel or a Hollywood film. As it stood before Bill Carney got involved, Abdul Ahad was the victims' next of kin. The grieving son and brother of the two civilian farmers who were killed while riding a motorcycle to their grape fields. If you recall, Abdul Ahad introduced himself to CID Special Agent Rios and Special Agent Mitchell while they were investigating the crime scene on the morning of July 7, 2012. Abdul Ahad was an information provider around whom the agents built significant elements of the criminal double-murder case. If true, if Abdul Ahad really was the Taliban leader of an IED cell operating in and around Strong Point Payenzai, this move of his—to approach army investigators proffering information—was the ultimate asymmetric-warfare trick. Like the fictional character Keyser Söze in the film *The Usual*

Suspects: a ruthless crime lord who was able to dupe law enforcement into thinking he was a nobody, give them misinformation, and get away.

"I truly believe that [these men] weren't out riding around picking grapes," Carney told the *Leavenworth* filmmakers.

John Maher knew how significant Carney's information was. He and the new defense team would build their case around it. Carney was now the team's biometrics expert witness.

"We filed for the petition for a new trial, [based on] *Brady v. Maryland* in 1963. A federal prosecutor, and a state prosecutor nowadays, has to disclose exonerating or mitigating evidence within its possession, which tends to negate guilt, reduce guilt, or mitigate punishment. None of that happened here," Maher said.

What Maher was saying was that if this biometric information was accurate, the army had violated the defendant's constitutional right to due process.

"The most glaring issue in Clinton Lorance's trial is [that] the Fifth Amendment due-process clause was violated. The [biometric] evidence was not only *not* disclosed, but it was withheld in the presence of an actual request," Maher told the filmmakers.

But there were problems with this storyline. How had Bill Carney accessed this biometric information—data that included the fingerprints and DNA of two murder victims, the witness Haji Karimullah, the detainee Mohammad Rahim, and the next of kin Abdul Ahad—without violating the Espionage Act of 1917?

In my researching and reporting for this book, Tom Bush, former head of the FBI's Criminal Justice Information Services Division, as well as numerous other subject matter experts, including William Vickers, who formerly advised the Deputy Undersecretary of Defense for Intelligence on biometrics capabilities, made clear to me that access to classified biometric data in military databases is strictly controlled.

CHAPTER NINETEEN

TRUE IDENTITIES

The Freedom of Information Act (FOIA) was created more than fifty years ago as a tool of transparency in a democratic society. It allows Americans to petition the government for information the government alleges should be kept secret. "An informed electorate is vital to the proper operation of a democracy," the courts ruled. The information that John Maher had presented, if true, was deeply troubling. If the U.S. Army really had withheld biometric evidence that might have negated guilt, reduced guilt, or mitigated punishment in a double-murder case against an American soldier, that forebode dark times in the age of biometric identification into which society is heading. Into a world where Identity Dominance—the distinct ability to separate friend from foe—is not just pursued by the U.S. government overseas, but inside the United States as well.

In November 2018, I began filing a series of FOIA requests to independently verify the information I'd been given. Maher and the defense team also shared with me the four biometric identification numbers that Bill Carney had produced using link analysis techniques he'd learned at the Afghanistan Captured Material Exploitation (ACME) lab at Bagram Airfield. As far as I knew, no journalist had yet taken an in-depth look at the Defense Department's fast-growing biometric identification system: how it was created, how data input worked, who controlled access, who guarded its guards. I'd read a scathing report by a principal adviser to the Secretary of Defense, calling the operational capabilities of ABIS "not operationally suitable" and "inadequate." Was the army covering up systemic failures within its multibillion-dollar system created to be "Manhattan Project–like" in its effect? The initial FOIA request I filed read as follows:

> November 11, 2018
> Defense Forensics & Biometrics Agency
> U.S. Army Office of the Provost Marshal General
>
> Freedom of Information Act (FOIA) request
> U.S. Army Freedom of Information Act Office
> Records Management and Declassification Agency
> 9301 Chapek Rd. Bldg 1458
> Fort Belvoir, VA 22060-5605
>
> Dear U.S. Army,
>
> As per the Freedom of Information Act (FOIA), 5 U.S.C. 552, I am requesting any and all records of the following four (4) Afghan males, each of whom was involved in IED incidents against US forces in Afghanistan, identified as such through DNA/fingerprinting as part of the Afghanistan Automated Biometrics Identification System (A-ABIS).
>
> These four individuals are involved in the general court-martial case of United States v. First Lieutenant

Clint A. Lorance, Army Docket Number 20130679, and Court of Appeals for the Armed Forces Docket Number 17-0599/AR, and this information falls within the public's right to know.

The names and biometric enrollment numbers for the four (4) Afghan males are:

1) Abdul AHAD B28JMUUYZ (Detention Facility Parwan, detainee)
2) GHAMAI Abdul HAQ B2JK9-B3R3 (killed on July 2, 2012)
3) Haji KARIMULLAH B2JK4-G7D7 (witness to July 2, 2012 killing[s])
4) Mohammad RAHIM B28JP-QWTY (wounded on July 2, 2012)

As a journalist and author [. . .], and with the information falling within the public's right to know, I request you waive fees associated with this search. Should you determine fees are required, please let me know.

Thank you.

Sincerely,

Annie Jacobsen

FOIA requests take an indeterminate amount of time to process, and so all I could do now, with regards to this part of reporting the story, was wait. The Lorance defense team continued to share numerous documents with me, including PowerPoint slide decks they created to get members of Congress interested in their quest and also to raise funds. Of note: the team shortened the name of one of the three motorcycle riders to a single-word name, for ease, they said. Ghamai Abdul Haq, one of the two men Clint Lorance was in Leavenworth prison for having murdered, was referred to as

"Ghamai" in all their material. The second man killed, Haji Mohammed Aslam, the village elder, had no known links to IEDs, Bill Carney said, and therefore had no biometric identification in the government database.

The way in which Carney linked the other four men to IEDs in Zhari and neighboring Panjwai District is central to the Defense Department construct of Identity Intelligence, or I2. "The I2 operations process results in [the] discovery of true identities," wrote the Joint Chiefs of Staff. The discipline of I2 uses data from multiple sources to discover identity by "connecting individuals to other persons, places, events, or materials, [and by] analyzing patterns of life." This battlefield intelligence tool, if you recall, is a foundation of activity-based intelligence, or ABI, which links individual people to criminal networks through geospatial data, including plotted maps.

The ABI methodology and I2 construct are also now being used in the United States. If an informed electorate is vital to the proper operation of a democracy, then it is worth understanding how this battlefield methodology is also being pursued by law enforcement here at home. One of the more suspect programs, born of Defense Department funding, promotes a controversial methodology called predictive policing.

Sometime around 2009, in Los Angeles, California, a small group of university professors started thinking about how gangs in Los Angeles and tribes in Afghanistan resembled one another. One member of the group, a criminology, law, and society professor named George E. Tita, expanded on this idea, he said, after attending a symposium in London, England, called the Defence Geospatial Intelligence Conference. "[I] listened to another group of presenters . . . talk about tribes and groups in Iraq and Afghanistan and realized that the motivations for violence, the reasons that people join these groups, whether they're gangs [or] insurgencies, are very similar," Tita recalled. "It was basically as if one could strike out 'Afghanistan,' replace it with 'Los Angeles.' Strike out 'tribes,' replace it with 'gangs.' That's really what brought me into this human terrain modeling arena," Tita said.

With funding from the U.S. Army Research Office, division of mathematics, and the National Science Foundation, Professor Tita and his university colleagues compared data from Iraq with data from Los Angeles gangs to determine how networks arise. Their goal, they wrote, was to use geospatial intelligence "to design and develop effective programs for reducing gang violence" in the United States. By conducting digital surveillance on small, 500-by-500-square-foot quadrants in Los Angeles, the professors created digital maps of human activity in areas where gang members were known to operate. "We [began] looking at the social and geographic spaces occupied by gang[s] as a group and also the individual gang members' patterns of association," Professor Tita explained. Unlike in Afghanistan, the Los Angeles Police Department could not require biometric captures from people on the ground. But the LAPD could, and did, conduct digital surveillance from overhead. The first tenet of activity-based intelligence is *You are what you do.* The professors' work used I2 and link analysis to presume *You are who you affiliate with.*

"We've been able to demonstrate that if you understand, at the group level, the relationships or the rivalries that link gangs together, that there may be certain nodes in a network that may have a larger impact on reducing crime and gun violence than if you were to pick some gang or set of gangs randomly. So the data is very important for helping one think about how to allocate resources," said Tita.

In doing this early research, another one of the professors, a UCLA anthropologist named P. Jeffrey Brantingham, had a eureka moment. He realized that with this kind of data, machines could be taught to predict crimes. To identify who is a criminal, and to stop them before they comment their next crime. Brantingham and a third member of the team, an assistant adjunct professor of mathematics named George Mohler, created an algorithm that would be the basis of this new law enforcement methodology. The professors trademarked their algorithm and founded a company called PredPol, short for predictive policing.

"A lot of human behavior can be explained with very simple mathematical models," Brantingham said.

"Criminals want to replicate their successes," Mohler told an audience at the Harvard Kennedy School. "They go back to similar locations, they repeat their crimes."

"Criminal offenders are essentially hunter-gatherers," Brantingham told *Science Daily*, "they forage for opportunities to commit crimes. The behaviors that a hunter-gatherer uses to choose a wildebeest versus a gazelle are the same calculations a criminal uses to choose [to steal] a Honda versus a Lexus."

In 2012, Brantingham and Mohler's Santa Cruz–based company became the first start-up to specialize in predictive policing. With $3.7 million in initial venture-capital funding, in just three years PredPol was being used by sixty police departments across the country. "Can we actually push policing to look into the future and make a reasonable prediction about the near term?" Brantingham wondered. "This is the type of research that is necessary to make that a reality." Research that can only be done by collecting vast amounts of data on individuals who have not yet committed a crime.

Andrew Guthrie Ferguson, professor of law at American University, has serious concerns about this kind of technology. "Predictive policing is only as good as the officers and analyzers who handle the data," he warns. PredPol promised to be the "holy grail of policing, preventing crime before it happens." In reality, Ferguson believes, PredPol demonstrates that "the technology has far outpaced any legal or political accountability." A study funded by the National Institute of Justice echoed this same concern. "[Humans] must distinguish intelligence from information, which determines what is and is not protected under privacy laws."

PredPol won't say how it handles data. Its software, like Palantir's, is proprietary. "We don't know how data is inputted or analyzed," Ferguson cautions, raising the questions: Who polices the analyzers? Who discerns information from intelligence? Who makes sure the algorithms don't have inherent biases or encoded mistakes? The story that Kevin H recounted involving the near-execution of the Payenzai farmer mistakenly identified by an algorithm to be the IED emplacer in the purple hat is one example of so much that can go wrong.

In 2011, *Time* named PredPol one of the 50 Best Inventions of 2011, alongside the electric car and devices to help paralyzed people walk. The tide of public opinion has since turned on predictive-policing technologies. The civil rights advocacy group Movement Alliance Project is among those opposed. "[These] pretrial risk assessment tools (RATs) can embed racial, gender, and economic bias into the factors they use, and the predictions they produce, but they frame those predictions as 'science.' RATs have the capacity to vastly over-inflate and over-predict risk."

PredPol uses algorithms to link criminals to other people who might be criminals, similar to the "Attack the Network" methodology pursued in Afghanistan. In PredPol's case, the algorithms have been written by mathematicians who see "criminal offenders [as] essentially hunter-gatherers [who] forage for opportunities to commit crimes." To understand how flawed algorithms can be, ABI subject matter expert Patrick Biltgen cites the British statistician George Box: "All models are wrong, but some are useful." An algorithm is a model, he says, a process or set of rules to be followed in a calculation or another problem-solving operation. "When you build a model, you have to make an assumption. The assumption is based on the modeler's view of reality." A 2018 exposé by Vice Motherboard revealed PredPol's algorithm was based on a questionable law enforcement strategy known as "broken windows" policing, one that states that visible indicators of crime foster other crimes. "PredPol takes an average of where arrests have already happened, and tells police to go back there," Motherboard reported.

Patrick Biltgen believes "humans must always remain in the mix," keeping overwatch on the intelligent machines. Biltgen, if you recall, was an engineer on DARPA's Autonomous Real-Time Ground Ubiquitous Surveillance Imaging System, or ARGUS-IS, built for the battlefield and said to include the most powerful surveillance camera system in the world. When DoD allowed PBS *Nova* to report on this 1.8-gigapixel surveillance drone, back in 2013, reporters learned that a single ARGUS-IS "was equivalent to 100 Predator drones hover[ing] over a medium-sized city." From 17,500 feet up, ARGUS-IS captured streaming video with enough resolution to see people on the ground, waving their arms, and to discern objects measuring just six inches in length. What, exactly, ARGUS had its eyes on, then or now, the Pentagon declined to say. The technology was mind-boggling. "It is important for the public to know that some of these capabilities exist," chief engineer Dr. Yiannis Antoniadis said. "This is the next generation of surveillance."

Despite its suspect algorithms, PredPol's predictive-policing technologies have been stealthily expanding across America. By 2020, police in more than 1,000 counties were using predictive-policing software systems. "With AI and criminal justice, the devil is in the data," says NYU law professor Vincent Southerland. PredPol's founders stand by their product and its methods. "PredPol is currently being used to help protect 1 out of every 33 people in the United States"—approximately 10 percent of Americans—a company spokesperson confirmed.

Waiting for a FOIA request to be answered can feel like *Waiting for Godot.* There's a chance that what you're waiting for will never arrive.

How had Bill Carney linked the four individuals—Ahad, Rahim, Karimullah, and Ghamai—to multiple IED explosions in Zhari and Panjwai Districts? Link analysis is based on forensic and biometric data, and I hoped information contained in additional documents I requested through FOIA would provide answers. At this point in my reporting, I understood the four men to be members of the same family, related by blood or marriage. The father (Haji Mohammed Aslam), a village elder in Payenzai, had no biometric links to IEDs, the defense team said, but his brother (Haji Karimullah) and son (Ghamai) did. Another son, Abdul Ahad, was the ringleader of the group, according to Carney, a most wanted terrorist marked for death. Abdul Ahad told army investigators he was the brother-in-law of Mohammad Rahim, the detainee shot in the arm.

In February 2019, I received 107 pages of declassified government documents pertaining to my FOIA request. There are two kinds of files on biometric identities culled from data in ABIS. Some are narrative, while others have pages and pages of data. For example, each electronic fingerprint is individually recorded as a list of numbers, letters, symbols, and words. Each reporting agency is similarly identified. Each encounter is dated. In the documents for Ghamai, there were pages of information pertaining to data captured by a BAT, a HIIDE, a SEEK, or a DNA swab in the battle space. By example, the right index finger of "Ghamai" appeared like this:

> Data file: U:\! DFBA STAFF\FOIA REQUESTS\2018-03\GHAMAI\020140413-035906-DDSOCOM95.EFT
> Record 4b: Fingerprint Image
> Right index
> Live-scan plain, Compressed image
> Width: 800, Height: 750, Compression Rate: 16:1, Offset: 43928, Length: 37147, IDC: 2 ANSI/NIST Image 1
> MD5 hash: [continued to next body part].

Because "Ghamai" apparently had an amputated finger, much of his fingerprint data also included numerical codes to account for the missing body part. While sifting through mind-numbing amounts of data, I came across an entry that I presumed, at first, to be an army reporting error. It was a date. In the document for Ghamai, next to a record of a DNA swab taken, I read:

Date of Arrest DOA 1 Apr 13, 2014

Ghamai was supposed to have been murdered by Clint Lorance on July 2, 2012. Now here he was, being arrested one year, nine months, and eleven days later. According to the declassified data in front of me, Ghamai was not dead. In fact, Ghamai had been detained by the army on "suspicion of terrorist activities" in southern Afghanistan four times since he was allegedly killed. According to the declassified documents I was reading, these arrests occurred on April 13, 2014; January 9, 2015; March 30, 2015; and May 12, 2017. Had there been a misidentification on the battlefield, or was it an ABIS error? Or something else?

I recalled a Defense Department biometric training presentation circa 2009 that warned: "Caution: Data is only as reliable as the person who collected and entered it!" The Biometrics Identity Management Agency that produced the slide presentation highlighted this warning in bright yellow.

"We don't know how data is inputted or analyzed," Professor Ferguson said of the big-data systems operated by the government. "If your inputs are distorted, you're going to get distorted outputs."

I wrote to FOIA officer David J. Meyer, in the Office of the Provost Marshal General at the Pentagon, asking him to look into the matter.

"I suspect you have sent me the wrong files on this aforementioned man identified as 'GHAMAI,'" I wrote. "Either that or Clint Lorance is in prison for having killed someone who, according to the attached U.S. Army records, is still alive. Please send me the correct A-ABIS files on GHAMAI Abdul HAQ, BID# B2JK9-B3R3, or provide me with an explanation as to why the U.S. Army biometrics database system says this man is still alive."

FOIA officer Dave Meyer's response was remarkably fast and succinct. "Ms. Jacobson [*sic*], We verified with our biometrics organization that the correct file based on the BID was provided."

What was going on? Bill Carney had gotten this BID from the Defense Department's ABIS database, Maher said. I reviewed Carney's sworn affidavit, one that was on file along with a Habeas Corpus petition to the commandant of the U.S. Disciplinary Barracks at Fort Leavenworth. There, Carney had certified that the testimony he provided was "based upon the standards and methods used by experts in the field of investigations generally and in Afghanistan concerning IED networks specifically."

I emailed John Maher telling him the FOIA documents said Ghamai was still alive. I asked to interview Carney immediately. Maher wrote back: "Annie—good morning. We sent the GHAMAI document to our biometric expert for his assessment, but, at first glance, as you noted, this is an important piece. Are you available for a call when we hear back from our biometric expert, so we can share with you the benefit of his analysis? Sincerely, John N. Maher."

Where was the flaw? The mystery had only deepened. The Defense Department's Identity Intelligence (I2) construct vowed to resolve identity. Its biometric identification systems were said to be the holy grail of positive identification. Yet in this situation, one that involved a double-murder conviction of an army officer, the true identity of one of the murder victims remained unresolved. Who was Ghamai? Was he a friend or foe, a farmer or a bomb maker? And was he even dead, or was he still alive?

Bill Carney remained tied up in Afghanistan, I was told by the defense team, doing important, classified government work. I'd have to wait.

During the war in Afghanistan, the discovery of true identities became a fixation of the Joint Chiefs of Staff. The Defense Department sought to assign to millions of individual Afghans a precise level of personalized risk. In military operations around Strong Point Payenzai, this was achieved in two ways. From overhead, cameras on Persistent Ground Surveillance Systems recorded pattern of life activities from persons of interest on the ground—for future use. On the ground, infantry soldiers like the paratroopers of First Platoon collected biometric information from everyone who lived in an area or simply passed through, to be stored in a classified database—for future use.

Because Afghanistan was an active war zone, the data from pattern of life activities could be used as evidence to support the killing of an individual person, usually by air strike. That person could be anonymous, meaning the Defense Department did not need to have positive ID on someone they killed, as long as that person's actions met a legal threshold, for example, the "three contacts with the ground" rule employed to determine that someone was a bomb emplacer. Infantry soldiers would then be sent to conduct a battle damage assessment, to get a positive identification off the dead person. The Defense Department would add and delete names off its classified kill or capture lists accordingly. Lists that all pulled data stored in ABIS.

While researching and reporting this book about what happened to First Platoon at Strong Point Payenzai, I was surprised to learn how powerful and invasive a surveillance platform the PGSS system really was. Very little information on how these systems worked in Afghanistan has been revealed in narrative form (most of it has been highly technical).

The United States lost the war in Afghanistan. The Taliban are in control of the country. A majority of the bomb makers detained and imprisoned based on forensic and biometric evidence have now been released by the Taliban. Clearly, America's rule-of-law programs in Afghanistan did not promote the rule of law. And this question naturally comes next: Is it a good idea that these same persistent surveillance systems have come home to America, where they are being used by law enforcement under the guise of rule of law?

In the United States, a person cannot be executed without arrest, trial, and conviction first. This is foundational to the rule of law, to the criminal justice system of law enforcement, courts, and corrections. The persistent ground surveillance systems currently being flown over American cities are there to loiter and record video in order to solve crimes at a future date. One of these companies, called Persistent Surveillance Systems, or PSS, is the brainchild of a retired air force officer, physicist, and MIT-trained astronautical engineer named Ross McNutt.

"Our goal is to help reduce murders in inner cities," Ross McNutt told me of his company's intentions. "To bring technological solutions to difficult social problems" through the rule of law. McNutt worked on the Pentagon's first live-feed, wide-area, persistent surveillance system called Angel Fire. After being refined at the Los Alamos National Laboratory in New Mexico (where Manhattan Project scientists built the atomic bomb) this surveillance system was flown over Fallujah, Iraq. After retiring from a twenty-year career with the military, McNutt created Persistent Surveillance Systems for law enforcement. Initially, he had trouble finding a police chief in America who was interested in the technology. His first substantial client was José Reyes Ferriz, the mayor of Ciudad Juárez, Mexico. The two men met at a security symposium in Miami, Florida, in 2009.

Flying a company Cessna over the dangerous Mexican border city, "We watched murders happen live," McNutt recalled to me. At the time, the Sinaloa and Juárez drug cartels had turned Ciudad Juárez into a city often referred to as "the murder capital of the world." Within the first hour that McNutt's surveillance cameras were flying overhead, the system captured two murders in real time, and another five murders shortly thereafter.

Despite the success in Mexico, it took years to get anyone in the United States interested, McNutt told me. Then, in 2012, while doing a demonstration test for Dayton police chief Richard Biehl, a compelling event occurred. Down on the ground, police officers received reports of a shooting and robbery at a small business called the Annex. It was the middle of the day. An employee triggered an alarm and the robber fled. Without overhead surveillance, the clue trail might have ended right there. Instead, PSS cameras captured what happened next.

After the thief left the store, he drove to a Subway and attempted to rob the sandwich shop. The shop owner pulled out a gun and the man fled. He drove to a Family Dollar store, spent several minutes inside, then left. On the way home, he stopped at Clark Gas Station to refuel, then went to a residential area. Several hours later, with the PSS cameras still on him, the man returned to the Family Dollar store and robbed it.

McNutt's surveillance system is made of an array of twelve commercially available Canon cameras mounted to a small Cessna aircraft that flies over a city in a circular path. The cameras focus on what's called an aim point, then project their lenses downward, in a trapezoidal footprint shape, onto the collection area on the ground. The imagery gets recorded onto hard drives, to later be analyzed and turned into an intelligence product by forensic investigators from McNutt's team.

But how to get positive ID on the suspected criminal? Unlike in Afghanistan where there is no infrastructure, Dayton, like every U.S. city, maintains a network of ground-based surveillance systems, including CCTV cameras and automated license plate readers, or

ALPRs. Data pulled from these sources helped Dayton police identify the robber and build the case that led to his conviction.

In 2014, *Washington Post* reporter Craig Timberg interviewed Dayton residents to learn what they thought. "There are an infinite number of surveillance technologies that would help solve crimes," said Joel Pruce, a University of Dayton postdoctoral fellow in human rights who opposes overhead surveillance. "You know where there's a lot less crime? There's a lot less crime in China."

In 2016, *Bloomberg Businessweek* reported PSS was conducting wide-area surveillance over Baltimore, including during the judge's delivery of the not-guilty verdict in the case of the Baltimore police officer charged with murder in the death of Freddie Gray. Baltimore residents had no idea they were being watched, because the city had not yet acknowledged that the program existed.

On April 9, 2020, the American Civil Liberties Union (ACLU) and the ACLU of Maryland filed a lawsuit against the Baltimore Police Department to challenge the constitutionality of deploying an even more advanced version of the surveillance system over the city. "[The] program would be the most wide-reaching surveillance dragnet ever employed in an American city," said ACLU attorney Brett Max Kaufman. "This technology is the equivalent to having a police officer follow you every time you leave the house." One of the plaintiffs, a Baltimore-based community activist and rapper named Kevin James, called the program Orwellian. "It is dark, sinister, and unjust," he says. "It is disquieting to know someone is watching you all the time."

But this is the whole point of the panopticon: to make people feel as if they are being observed. "A single camera mounted atop the Washington Monument could deter crime all around the Mall," McNutt believes. Proponents of overhead surveillance say it reduces crime because the fear of being watched has the same impact as actually being watched. That if you have nothing to hide, you have nothing to fear. But this mentality overlooks the fact that today's digital panopticon is light-years in conception from the physical building Jeremy Bentham imagined in the eighteenth century as a system of control. The panopticon of the twenty-first century has grown in too many secret ways to reasonably accept it can be trusted as a fair and just component of rule of law.

In September 2001, terrorist attacks carried out by nineteen individuals caught the U.S. national security apparatus entirely off guard. Asked how not to let such a horror happen again, the Defense Science Board told the secretary of defense, "The [global war on terrorism] cannot be won without a 'Manhattan Project'–like tagging, tracking, and locating program for national security threats." Since then, tens of millions of people overseas and in the United States have been identified, or tagged, and their biometric data placed into a database for positive identification sometime in the future for a crime they have yet to commit.

In the United States, it used to be that a person had to break the law to wind up in a criminal database. But the prerequisite of prior criminal activity is quietly being undone. This is what Justice Scalia warned of in *Maryland v. King,* when he alerted society to the perils of a genetic panopticon. A person's unique DNA profile—also called a DNA fingerprint—can be obtained entirely without their knowledge or consent because DNA gets left behind everywhere, by everyone, all the time. For a glimpse at the dangers that lie ahead, we can look to a recent case involving New York City detectives who secretly took a DNA sample from a twelve-year-old boy.

CHAPTER TWENTY

THE GENETIC PANOPTICON

In March 2018, New York City detectives were questioning a young boy, whom they had arrested in a felony crime investigation, when they offered the child a soda from McDonald's. The boy accepted and after he finished the drink, one of the detectives surreptitiously took the cup away. Donning rubber gloves, the law enforcement officer removed the straw from the cup, bagged and tagged the item, and sent it to the crime lab. There, technicians extracted micrograms of the child's DNA left behind by his lips. Their goal was to create the boy's genetic fingerprint from which they could try to obtain a crime scene match-hit. A person's genetic fingerprint is the same in every cell and retains the same unique distinctiveness across a person's lifetime.

The child's DNA profile was uploaded into a database called the Local DNA Index System. The original samples were kept stored by the New York City's Office of Chief Medical Examiner.

When *New York Times* reporters Jan Ransom and Ashley Southall broke the story in 2019, readers expressed shock. How could police officers take a person's DNA—in this case, the DNA of a child—without that person's or their guardian's knowledge or consent? As it turned out, the boy's DNA profile did not match evidence from the crime scene and felony charges against him were eventually dropped. But as Ransom and Southall reported, it took the boy's mother more than a year to get her child's genetic profile removed from the government's criminal database.

As Fourth Amendment rights regarding search, seizure, and the right to privacy are being debated in the courts, advances in forensic DNA are moving forward at science fiction–like speed. How did we get here and where is society headed next? To consider how DNA impacted the final chapter in the saga of First Platoon is the canary-in-the-coal-mine tale.

Ninety years after the first bloody fingerprint was used to convict a criminal in a court of law, in Argentina in 1892, forensic science experienced another revolutionary leap forward with the discovery of genetic fingerprinting. It was 1984, and British scientist Alec Jeffreys was working with DNA in a laboratory he ran for the Department of Genetics at the University of Leicester in England when he made history.

"In science it is unusual to have such a eureka moment," said Jeffreys, who was knighted by the queen of England for his contributions in forensic DNA. Genetic fingerprinting is a multi-step scientific process that begins with extracting DNA from a biological sample. Then, using enzymes, the sample is cut into fragments, separated according to size, and treated with a radioactive probe, allowing certain details to be captured on X-ray film. The result is a pattern of more than thirty stripes, or bands, of DNA that resemble a bar code.

Jeffreys's initial focus was finding a method to resolve paternity and immigration disputes using genetics. But a series of unsolved murders in nearby Narborough Village catapulted the concept of using genetic fingerprints to solve crimes into the public domain.

Not far from Jeffreys's science lab, a serial rapist and killer was on the loose. Two young women were dead and local law enforcement had no leads. In an effort to solve these grisly crimes, police set up the first mass DNA screening in history. Thousands of local men were asked to give blood samples for DNA comparison against sperm cells left by the killer on the victims' bodies. When, at a local pub, a baker named Ian Kelley was overheard talking about how a colleague named Colin Pitchfork had convinced him to give a blood sample on his behalf, Pitchfork became the prime suspect. With probable cause established, police obtained a blood sample from Pitchfork, whose genetic fingerprint matched the crime scene evidence. This was the world's first serial-murder case solved using DNA evidence in a court of law.

Across the Atlantic, Jeffreys's work interested the U.S. Department of Defense. In 1991 the Armed Forces DNA Identification Laboratory was established as the military's first forensic DNA-testing laboratory. In Rockville, Maryland, DoD's morticians and technicians worked to identify the repatriated remains of soldiers, marines, and airmen. These were mostly human bones and teeth found scattered across former battlefields, like in Vietnam. Match samples were taken from next of kin, and learning the identity of the remains was a long, laborious process. The Defense Department had determination and resources. And yet, lab work, analysis, and peer review took, on average, twenty-four months.

The following year, at the behest of the secretary of defense, the DoD began expanding its database from dead soldiers to its living service members, requiring that blood specimens be taken from all troops returning home from Desert Storm. "Two drops of blood" on a paper card "that form stains about the size of fifty-cent pieces," air force doctor Colonel Vernon Armbrustmacher explained, when asked to provide details. The military kept its DNA samples stored in the basement of its laboratory, each card in a vacuum-sealed envelope, inside freezers running at negative 4 degrees Fahrenheit. Five years later, the Department of Defense had DNA specimens from 1.2 million service members. By the turn of the century, every person in uniform had a DNA sample on file. On March 29, 2019, the U.S. military ceremoniously processed its eight millionth DNA reference card—a genetic catalog of every service member since 1992.

As far as DNA samples from foreign nationals were concerned, there was no official military program until February 2001, when a small group of military and intelligence community partners proposed the creation of a highly classified, searchable database to house the genetic fingerprints of the world's most wanted men. Called Black Helix, the program has only ever been mentioned in a single line in official literature, in a 2007 report by the Defense Science Board Task Force on Defense Biometrics. Black Helix was created as "a secure repository and interactive database, which will focus on archiving, retrieving, and interpreting bio-molecular data for the identification and tracking of terrorist suspects."

To the Pentagon's science advisers, DNA had an appeal above and beyond the three biometrics the military was focused on at the time (that is, fingerprints, iris scans, and facial images). "Residual traces of DNA can frequently be found 'at the scene,'" the Defense Science Board advisers noted. Further, "we do not require the subject's cooperation or even awareness to 'enroll' said individual." On the downside, DNA was "neither quick nor cheap to process." Certainly not in 2007. But all that was about to change, starting with the untenable backlog of DNA samples being captured from foreign fighters held in U.S. prisons across Iraq and Afghanistan.

The Pentagon set up a new DNA database, called the Joint Federal Agencies Intelligence DNA Database, which in 2007 contained more than 15,000 DNA profiles, but with a queue of 30,000 new samples waiting to be analyzed. In an attempt to speed up these efforts, the Defense Department looked to its FBI partners for help. But the bureau's lab at Quantico was equally overwhelmed, manually processing just two samples every three days, while receiving 9,000 samples a year. DNA lab work was delicate and time-consuming, most notably the part involving peer review, the process whereby one human checks, and rechecks, another human's work for possible errors. It was speed, or lack thereof, that was holding back the widespread use of DNA as a combat biometric. Comparatively it was especially slow considering electronic fingerprint files took just seconds for a match-hit.

Along came a Harvard-educated geneticist and former pediatrician named Dr. Richard Selden with a seminal idea. He was going to bring DNA testing out of the laboratory and onto the battlefield. He was going to invent a small, rugged, automated machine that could deliver laboratory-quality DNA results with minimal human effort in *less than two hours.* That was his goal. It was precisely what the Defense Department was looking for.

Like many inventors, it was a science fiction novel that inspired young Richard Selden to imagine and innovate. For Selden, the story was *Joshua, Son of None,* a fictional tale, published in 1973, about a medical doctor who cloned President Kennedy by stealing a sample of the slain president's DNA from the hospital in Dallas. The doctor's goal was to guide the replicant JFK back into the White House to become president once again. Instead, in the story, fate and circumstance proved more powerful than science and the cloned JFK was assassinated once again.

"You can't stop the future from happening," Dr. Selden said to me when we first met at the International Symposium on Human Identification in Palm Springs, California, in 2019. But the story of *Joshua, Son of None* planted a seed in Richard Selden's childhood mind.

"It gave me ideas," he said.

When Selden reflects upon his journey into the world of forensic

DNA, it bears the context of decades of experience in the field. Like former FBI CJIS Division director Tom Bush, who was also born in the 1950s, Dr. Selden straddles a mostly preautomated world. The 1970s and '80s were frustrating to him for exactly this reason, Selden recalled. He would look at the great advances in information technology, in computer technology, and lament how little progress had been made with DNA. The way Selden saw it, "Ever since John von Neumann invented the ENIAC computer, in a basement [in Princeton] in the late 1940s," information technology took off. "Whereas IT thinkers had gotten out of the basements, we DNA thinkers, the molecular microbiologists, we were still living in ivory towers." Their work was rarefied, Selden learned as a graduate student at Harvard, mapping plant genes that had never been mapped before. Conventional wisdom was "day-to-day DNA does not impact the lives of most people." Selden disagreed. "DNA should be making a difference to everyone, on a day-to-day basis," he believed.

While working as a pediatric emergency room resident in Boston, in the 1980s, Selden had his own eureka moment. Infectious diseases were a top killer of children, he recalled, as was "the poorly directed use of antibiotics," which was often based on trial and error. One night, waiting for test results for a child at the edge of death, he realized the simple truth: "Time is against doctors." Selden sought grant money to work on this problem, but found none. "Everything is requirement-driven," he found.

Nearly two decades passed. Then, in 2005, Selden won a grant from the National Institute of Justice to explore how to speed up genetic-testing results for law enforcement. From contacts in that world, he learned the Defense Department was interested in funding the kind of work he was interested in pursuing. Not for identifying antibiotic-resistant genes in children per se, but for more rapidly identifying DNA left behind on IED components on the battlefield.

"What I found, in talking to people in Homeland Security and the military," said Selden, "was that the idea of getting DNA results quickly was actually a critical part of national security. What military person can wait twenty-four months to get a friend-or-foe test back?"

With seed money from the Defense Department, Selden and a team from his company, called ANDE, got to work making the process accurate, durable, and fast. The work was requirement-driven, with significant hurdles to overcome. DNA had to be purified from samples collected by buccal swabs, from Q-tips, not needles. A lightweight system needed to be developed to quickly extract junk DNA from usable DNA; there were no subzero freezers in the field; specimens had to be freeze-dried. The ruggedized, portable DNA-testing system needed to be the size of a large microwave, able to withstand shock and vibration in the belly of a C-130 aircraft, sustain a fall out of a truck. It had to have no obvious parts to break, which meant no On/Off switch. "[DoD] said a sixth-grader needed to be able to operate it," Selden recalled, "an average sixth-grader," not a child prodigy.

The work took years. The team was near completion in 2012 when the FBI announced it was changing standards to make the science more precise, "which was good, but it set us back." There were instrument changes to make; the team pressed on. Finally, in 2014, Selden and his team completed their invention: a portable, rapid DNA identification system that delivered laboratory-quality results in just ninety-four minutes. DNA was now positioned to take the lead as a biometric favored by the Defense Department in a war zone. A ninety-four-minute DNA test result was within the timeframe that detainees could be held, before the rules of engagement required release or longer-term detainment. "DNA is by far the most accurate biometric," Selden clarifies. "A fingerprint might be 92% accurate. But DNA is 99.9999999999—that is ten nines after [the decimal point]—percent accurate."

But by 2014, the Pentagon was pulling its forces out of Afghanistan, which meant Selden's rapid DNA systems would get little use on the battlefield. Instead this technology would be shared with the DoD's federal law enforcement partners in America. But first, it was a nonclassified system of DNA link analysis that would take forensic DNA in a radical new direction here at home—to a place even Dr. Selden says he could not have foreseen.

In 2010, two retired businessmen in Florida started a personal genomics website called GEDmatch. The goal of the open-source database, its founders said, was to help adoptees locate birth parents, help families locate missing family members, and help researchers fill in family trees. The site was free with registration. Users uploaded raw DNA files from commercial DNA-testing companies like 23andMe. Hundreds of thousands of amateur researchers and genealogists became involved, learning whom they might be related to, with the ease of an online database. But without any of the users, or anyone at GEDmatch, knowing, law enforcement officers were also using the database, covertly working to get information about the relatives of serial killers and violent criminals, through a process called forensic genealogy.

One of these cases made national news. California law enforcement officers working to learn the identity of an anonymous serial killer known only as the Golden State Killer uploaded the suspect's DNA profile to GEDmatch. This DNA profile came from a rape kit that had been in police possession for years. Using the GEDmatch database, investigators determined a pool of possible people related to the Golden State Killer, then used detective work to rule out certain family members as suspects. Their search narrowed in on a retired police officer named Joseph James DeAngelo. In 2018, with probable cause established through forensic genealogy, the detectives obtained a DNA specimen from DeAngelo's garbage and got a DNA match-hit that led to his arrest.

When the public learned the Golden State Killer had been identified and captured through familial DNA files located in GEDmatch's online database, the story kicked off a national debate.

Shouldn't DNA data be guarded? Wasn't this invasion of privacy? As it turned out, there were no existent laws protecting the DNA information of the people who had voluntarily entered their data into GEDmatch's database. Familial DNA linking meant users' unwitting relatives were equally unprotected, be they serial killers or law-abiding citizens. The regulations have since changed, but the data was already out there. Data, once made public, can get copied, captured, and amassed ad infinitum.

A look at China presents several worst-case scenarios of all that can go wrong, most notably when the state decides certain groups of people are a national security threat. In 2016, the Chinese government unveiled a state-mandated program called Physicals for All, and China began to identify and catalog all citizens of Uighur descent through mandatory DNA specimen captures. Uighurs are a Muslim minority, native to the Xinjiang Uighur Autonomous Region in northwest China. "Segments of the population have resisted Beijing's rule," says subject matter expert Dr. Anna Hayes. "Many refuse to speak Mandarin, while others campaign for independence," hardly considered criminal offenses in the West. But the Chinese Communist Party sees any form of criticism as a national security threat. Between 800,000 and 2 million Uighurs are believed to have now been detained, against their will, in secretive camps, under the guise of a reeducation program. The following year, in 2017, China's official news agency, Xinhua, reported that in addition to the DNA samples, the Physicals for All program netted biometrics on 36 million Uighur Chinese—including iris scans, facial images, voice prints, and more.

"The mandatory data-banking of a whole population's biodata, including DNA, is a gross violation of international human rights norms," said Sophie Richardson, Human Rights Watch, China director. That the program echoed the Defense Department's program in Afghanistan was ignored.

Technology from American companies is used in China's dragnet, including equipment made by the Massachusetts-based Thermo Fisher Scientific. And at least one known invasive offshoot of this program—outside China, here in America—has been made public. To locate Uighur family members who might be trying to hide their ethnicity from the state, China's Ministry of Public Security uses forensic genealogy. From Yale School of Medicine emeritus professor and geneticist Dr. Kenneth Kidd, the Chinese government purchased the genetic profiles of 2,143 Uighurs that Dr. Kidd had collected, for what he said was academic research. "I had thought we were sharing samples for collaborative research," Kidd told a reporter with the *New York Times*.

The science continues to advance into areas previously unimaginable. In a laboratory in Deerfield Beach, Florida, geneticists with DNA Labs International have pushed the science even further. Using a trademarked proprietary software called STRmix, they have developed ways to identify what a criminal might look like, from micrograms of captured DNA. This process is called probabilistic genotyping. Starting in 2017, probabilistic genotyping–based predictions about human traits such as eye, hair, and skin color have been used in court, similar to the way in which sketch artists create drawings of what a criminal might look like. Advances to build models of a person's probable identity from the gene up are under way.

"Eventually, testifying for cases involving probabilistic genotyping will be the new normal," says DNA Labs International spokesperson Samantha Wandzek.

As for Selden's rapid DNA system, once it got put to use by law enforcement across America, the technology began making headlines—mostly positive at first. After a wildfire in Paradise, California, killed eighty-five people in the fall of 2018, Selden's instrument was used by the Sacramento County coroner to positively identify all but two sets of skeletal remains. The following year, the system helped the U.S. Coast Guard again positively identify the otherwise "unrecognizable" bodies of all thirty-four victims incinerated in the *Conception* dive-boat fire, off the coast of Santa Cruz Island. In both situations, rapid DNA tests were perceived as working for the public good, bringing closure to distraught family members in times of great crises.

Then, in the fall of 2019, the *Washington Post* broke the story that federal officials were using Selden's system to test DNA samples taken from migrant families arriving at the U.S. border, as part of an "unprecedented" pilot program, and the concept of rapid DNA took on darker overtones. The players included the Department of Homeland Security, Immigration and Customs Enforcement, Health and Human Services, and Palantir Technologies—the data aggregation company that helps the Defense Department and its intelligence community partners identify which human targets can be tagged, tracked, located, and legally killed overseas.

It would take another year, and a global pandemic, to reveal just how draconian a surveillance network was being constructed here in the United States, and with the help of Palantir. A scenario PGSS operator Kevin H warned of earlier, in an interview for this book.

"The military application of Palantir is awesome," Kevin H told me in 2019, and warned of "other moves afoot" that suggested the government would employ Palantir to track large groups of people in the United States. When we met in person in the fall of 2020, Kevin H had this to add: "What Palantir is capable of is straight-up Big Brother. People should pay attention. For real."

Months had passed since I first requested a call with biometrics expert Bill Carney. The reason for the delay was always the same. Carney was working on classified government programs, John Maher said. He shared with me a letter written by eight Republican members of Congress, to President Trump, imploring the commander in chief to intervene on Clint Lorance's behalf. This letter, written on official

"Congress of the United States" letterhead, contained a mysterious detail involving DNA. "Dear President Trump," it read. "Fingerprint and DNA evidence following the attack identified the assailants as known enemy combatants, despite the prosecution claiming their status as civilian."

That the congressmen warranted to the president that the DNA evidence had been produced "following the attack" was troubling. How could this be? No one from First Platoon had taken samples of the men's DNA after they were killed on July 2, 2012—not according to the declassified CID report. So where had the DNA come from? Had Carney accessed the Defense Department's DNA database to discover that army investigators really had exhumed the bodies after all? If military investigators had taken DNA samples from the men killed and withheld this evidence at trial—that would be stunning.

Then, the defense team dropped another bombshell. This time, Carney's link analysis had led to an even more gut-wrenching reveal. "Staff Sergeant Israel P. Nuanes [was] killed by the motorcycle rider Ghamai," co-counsel Don Brown wrote in a self-published book, which he discussed on Fox News. Again, the specifics were uncanny: "In the biometrics database, the explosion that killed the American soldier has an event number," he wrote. "In Clint's case, IED Event #12/1229 designates the bomb explosion tied to the motorcycle rider Ghamai . . . Ghamai's DNA was found on [the] bomb."

It was mind-boggling. If Ghamai's bomb really did kill Staff Sgt. Israel Nuanes, this was yet another extraordinary plot twist. To the congressmen and the defense team, Nuanes might have been little more than a name—another tragic casualty of war. But to the soldiers of First Platoon, he was their EOD tech.

It was time for me to travel to Georgia, to run this scenario by Samuel Walley and Daniel Williams, and to discuss the perplexing reality of war. Even if one of the Afghan men really had built a bomb that killed Nuanes, the army's rules of engagement were such that Lorance had still unlawfully ordered them killed. As for the paratroopers of First Platoon, despite being vilified and ostracized by many members of the military community, they remained steadfast in their conviction—as individuals and as a platoon—that what Lieutenant Lorance did on the battlefield was unlawful and he was justly convicted of the crimes.

"What [Lorance] did dishonors every soldier who serves honorably," Captain Swanson told me. "You follow ROEs whether you agree with them or not."

"War is hard," Sergeant Williams told *New York Times* reporter Dave Philipps in 2015, "there is collateral damage, I get that—I've got my own stories. That's not what this was; this was straight murder."

The odds of Ghamai's bomb killing Nuanes were astounding. Almost too incredible to comprehend. Talking to Walley and Williams, in person, would be valuable. Too many biometric data points were not adding up.

CHAPTER TWENTY-ONE

THE COURT OF PUBLIC OPINION

In the summer of 2019, I traveled to Georgia to meet Samuel Walley in person. Our previous interviews had been on the telephone. When we meet, Samuel Walley is twenty-seven years old, six feet tall. He has green eyes, sandy brown hair, weighs 178 pounds, wears a neatly trimmed beard. He is fit and unflinching, practices jujitsu, rides jet-skis on nearby Georgia lakes. Walley is also missing his right leg from just above the knee down, and his left arm is missing below the curve of his elbow. The limb abruptly ends in a surgical twist of skin.

Here in the parking lot in Gainesville, Georgia, where we walk, Walley observes me, and his surroundings, with intensity and focus. He walks in almost perfect balance, having mastered the use of a titanium leg prosthetic. Part of his left leg, the one of flesh and bone, is dented and scarred with burn marks and skin grafts—the tibia was once part of someone else's body. Walley underwent a successful cadaver transplant surgery in 2013. Walley wears a baseball cap, shorts, and a T-shirt made of material that breathes.

"Amputees have body-heat-regulating issues," he says, and explains thermoregulation complications, the short version of which is: "Your brain thinks your limbs are still there."

Before the war, Walley had two atomic bombs tattooed on his left arm, and the word "Chaos" written beside, his nickname from high school. "Chaos" got blasted off by the IED on June 6, 2012, but I can see what is left of the inked drawing of the Fat Man bomb, born of the Manhattan Project, and how the image has now warped.

Seven years have passed since First Platoon was in Afghanistan, since that period of time when many lives were forever changed by a month-long series of events in Zhari District. Upon return to the United States, the platoon continued to suffer grave losses. The first to die was Mark Kerner. In a military hospital in Germany, when his body was scanned for shrapnel left over from the IED blast that tore away a portion of his buttocks and backside, doctors found cancer.

"He was recovering nicely for a while," remembered Todd Fitzgerald, one of his closest friends, "everything going well, in remission, until the cancer came back." Mark Kerner died in February 2015, at the age of twenty-four.

Matthew Hanes died six months later, in August 2015. Paralyzed and wheelchair-bound, he remained optimistic and enthusiastic until the end. Hanes's death, from a blood clot, ripped everyone's heart out, James Twist recalled. A twenty-one-gun salute at the Mount Rose cemetery in York, Pennsylvania, honored Hanes's life and service. Forty paratroopers from the 82nd Airborne were there. Matthew Hanes was twenty-four years old.

"After Hanes died, I started to feel crazy," Walley told me. "The nightmares worsened." He drank copious amounts of alcohol in an attempt to alleviate the pain. Walley lived alone in a small apartment in Georgia and spent much of his time cleaning his large collection of firearms, he says. "Everyone in the complex saw me as some crazy veteran who partied a lot and took pills for PTSD. Most of them were scared of me."

One night, home alone, cleaning his guns, things came to a crossroads.

"With a Glock, you have to pull the trigger to clean it," he clarifies. By accident he fired the weapon into the ceiling. Neighbors called the police. "I had on night vision and body armor. They took all that into consideration. A SWAT team arrived."

After law enforcement surrounded the building, he was ordered to come out with his hands raised. Walley complied. He wore a prosthetic arm at the time. The old kind, with a hook hand.

"Get on your knees!" one of the police officers shouted.

"I only have one knee," he shouted back. "If I get on it, I'm going to fall on my head."

He was drunk. Cameras were rolling. He remembers one of the police officers approaching him cautiously. A woman.

"She cuffed my prosthetic hook hand. It fell off."

The police officers began arguing.

"Why didn't you cuff him right?" one of them shouted.

"He's only got one damn hand," the officer shouted back.

The incident resulted in Walley being charged with two counts of felony aggravated assault, reckless conduct, and making a terrorist threat.

"I was looking at forty years."

After weeks of heavy drinking, Walley decided to kill himself.

"I decided to slit my throat with a knife."

He drank a bottle of whiskey. Just as he was about to do it, his father showed up unannounced.

"The door was locked. I had a key. I walked into the bathroom and there he was, out of his wheelchair on the floor," Walley's dad, Kelly, told me. "He didn't have his prosthetic on, just the nub. He had a Ka-Bar knife to his neck. I said, 'Son, please don't do it.' He said, 'I can't handle it any more.' I just sat there. He put the knife down. He talked about what went on over in Afghanistan. It was just too much for him to handle." Kelly had the insight to realize the wise move was to have his son call one of his brothers from the platoon.

"I called Haggard," Walley remembered. "Haggard said, 'Come now.'"

Walley's dad drove. Eight hours later, they were at Haggard's apartment in Ohio.

"We were sitting there together when we got a telephone call that Ruhl was dead," Walley remembers. Another incomprehensible blow. Jarred Ruhl had been at a party in a friend's garage in Fort Wayne, Indiana. An argument broke out. Ruhl had a gun. He'd been struggling with PTSD.

"Ruhl's brother tried to intervene and take the gun away," Walley says. The firearm went off and Ruhl was shot in the abdomen. He was rushed to the hospital, where he died. Jarred Ruhl was twenty-four years old.

Walley had a realization at the funeral. "Staring into Ruhl's open casket I realized if I didn't change my ways, my funeral would be next. I entered a veterans' rehabilitation program. Started helping other vets who'd lost limbs in combat."

Walley turned it all around. Became focused on attending college, and began to excel. Started an internship for Congressman Doug Collins. Became an inspiration among his platoon mates. For others, the healing process would also begin. Then, in January 2017, the narrative that Clint Lorance had been unjustly convicted of murder returned to national news like a house on fire.

Six days after assuming the office of the president, Donald Trump had invited Fox News host Sean Hannity to the White House for the new president's first sit-down interview.

"Let me ask you about the power of the pardon, which is absolute for a president," Hannity said to Trump. "One night, I know you were watching my show. Clint Lorance got thirty [*sic*] years [in prison]. He was doing his job protecting his team in Afghanistan."

President Trump nodded in acknowledgment. He appeared to be familiar with the case.

"We're looking at a few of them," the president told Sean Hannity.

Ever since John Maher had taken the lead on the Clint Lorance case, he and a team of defense attorneys had filed numerous appeals, culminating in a Writ of Habeas Corpus to the commandant of Leavenworth himself. The army rejected each appeal on grounds that it lacked merit. When Donald Trump was elected president, their strategy changed.

"I realized: there's an opportunity here," Maher said. The president's interest was a financial gold mine. Maher teamed up with a nonprofit organization called United American Patriots. He had been donating much of his time, he said. Now UAP could pay his substantial legal fees.

"Most of our donations are under forty dollars and are from red-blooded Americans who think putting combat veterans in prison is dead wrong," the organization's founder Major Bill Donahue told me. The United American Patriots have a singular purpose, Donahue said, and it is also made clear on the nonprofit's website: "to provide legal defense for American service members unjustly convicted and wrongfully imprisoned on war crimes charges."

Sean Hannity's 2017 interview with President Trump brought national attention to Clint Lorance's case. For United American Patriots, the president's remarks resulted in a flood of money, Maher recalled. He traveled to the U.S. Disciplinary Barracks at Fort Leavenworth, Kansas, to share the good news with his client. The encounter was filmed for the Starz documentary *Leavenworth.* Clint Lorance wore a prisoner's brown jumpsuit and name tag. John Maher wore a suit.

"That night, the phone did not stop [ringing at UAP]," Maher told Lorance, "probably until the next day at five o'clock."

"Really?" Lorance asked.

"Yeah. [Rang] all the way through the night."

From here on out, Maher and his team would use politics to frame, and to amplify, their cause. If you were pro-Trump, you should be pro–Clint Lorance. Emboldened by the national attention, the following year, the defense team appealed to President Trump himself. The letter was also addressed to Lawrence Kupers, the acting pardon attorney in the Justice Department, and Mark Esper, the secretary of the army.

The documentation supporting the defense team's extraordinary fingerprint and DNA claim was now on the desk of the president of the United States.

"Dear Mr. President," the letter read, "we now know by use of biometric evidence (fingerprints and DNA) that the males of apparent Afghan descent that Clint's Platoon shot and killed were not civilians as the prosecution claimed, rather, they left their fingerprints and DNA on improvised explosive device [*sic*] (IEDs) at GRID locations where American paratroopers were blown up and killed."

This was a stunning claim to make to the commander in chief of the U.S. Armed Forces. There is no one higher in the pyramid of power, in the chain of command, than the president of the United States. In the two-page letter, the defense also singled out Ghamai and Haji Karimullah by name. After reminding the president that Clint Lorance had been "sent to federal prison for 20 years," for murder and other crimes, they wrote: "What went undisclosed, however was that Haji Karimullah, whom Clint stands convicted of attempting to murder, left his fingerprints on IEDs at a GRID coordinate where US paratroopers had been killed. Ghamai, a rider who was shot and killed, who Clint stands convicted of murdering, also left his fingerprints on IEDs."

And yet the information I obtained through FOIA requests contradicted this claim. Based on what I knew, the murder victim

Ghamai wasn't even dead. What biometric data did the defense team have that held such sway?

It all seemed very convoluted. Reporting a nonfiction story requires scouring source material again and again, in order to become familiar with facts and discern possible leads. Journalism often gets compared to detective work. One of the reasons I traveled to Georgia to meet Samuel Walley and Daniel Williams in person was to go over the grid locations that the defense team referenced in its letter to the president. When discussing the death of Staff Sergeant Nuanes by Ghamai's IED, co-counsel Don Brown had identified the blast by an IED event number. Through the FOIA, I'd confirmed that IED event, but I also needed to confirm its precise location. Using the military grid reference system, I plotted this geocoordinate on a map to bring with me to Georgia. Walley and Williams were two people who had unique, firsthand knowledge about where EOD tech Israel Nuanes was actually killed. I showed the map to them.

It was a hot, humid summer day in Georgia. We sat in the lobby of the hotel where I was staying, drinking iced tea from the hotel's hospitality table.

Walley looked at the map, with a military geocoordinate for IED Event #12/1229 clearly marked.

"That's not where Nuanes was killed," Walley said firmly.

"Nuanes was killed way over here," Williams said, pointing to a location on the map approximately fifteen miles west. In an entirely different district.

I said, "Clint Lorance's defense team insists this grid coordinate is where Staff Sergeant Nuanes was killed by an IED. They've made this claim to President Trump."

"Lorance's defense team is full of shit," Williams said with certainty.

"I know where Nuanes was killed," Walley confirmed.

Williams picked up my pen and marked the true location with an *X*.

•• •• ••

After I returned home from Georgia, I hired scholar Muhammad Sajjad, PhD, of the Institute of Geology, University of the Punjab, in Lahore, Pakistan, to assist me in making maps of the area around Strong Point Payenzai. Many maps of the area published in English-language books are inconsistent. For example, Zhari District village locations in the authorized biography of General Petraeus differ from ones published in monographs by scholars with the Institute of War think tank. The best nonmilitary mapmakers, I learned from former GEOINT officers, are often regionally based academics who are also familiar with local languages. Scholars match data sets against the names of small villages, using local language filters. I provided Muhammad Sajjad with the military coordinates I'd obtained from declassified army documents, and he got to work. As his drafts came back to me, I shared them with several of First Platoon's former paratroopers, in order to establish locations referred to in this narrative. The most enthusiastic person in this effort was James Twist.

Since returning home from the war, Twist, too, experienced a transformation.

"At first life was terrible," he told me. "I spent too much time drinking. Not partying, drinking. I struggled." Then, after four years working what felt to him like a dead-end job at Costco, Twist had an epiphany: "I'm cutting meat. Stocking hamburgers. It's a menial job. You come to work, you screw off. Repeat. That's when it hit me. I had no purpose for being there. I needed meaning in my life."

Twist decided to become a law enforcement officer. He entered the Michigan State Police Training Academy in 2017 and graduated in July of the following year, Trooper #1615. In the eleven months we spent interviewing for this book, he was employed as a Michigan State Trooper. He and his wife, Emalyn, had three children. Two boys and a girl, ages one, three, and five. Life was great, Twist said time and again. Missing the camaraderie of the army, he'd joined the Army Reserve and served with the 321st Psychological Operations Battalion in Grand Rapids. Age twenty-seven, he was a family man now, he told me. During our interviews, I would sometimes hear him chopping up celery and carrot sticks for the kids' lunchboxes. "To go with the hummus," Twist joked.

James Twist enjoyed reviewing drafts of the maps I was working on, he said. I would send him electronic files. He'd send them back, marked up in colorful electronic ink: *this happened here, that happened there, you're off by two grape fields*—that kind of detail.

"Keep sending," he wrote, "I love this stuff."

In late summer 2019, another group of FOIA documents arrived, with information that further contradicted the defense team's claims. The data in this new set of FOIA documents stated that the man they called Karimullah—the third motorcycle rider, the man dressed in all white—served as a member of the Afghan Local Police. His name was Karimullah Abdul Karim, not Haji Karimullah. The IED on which his fingerprints were discovered was a misidentification, the declassified document clearly stated. "Based on the subject being part of the Afghan Local Police (ALP) it is most likely this match is due to contamination," one military analyst wrote. Elsewhere in the file, an FBI examiner with the Terrorist Explosive Device Analytical Center agreed with this assessment: "Analyst comment: It is likely that Karim [i.e., Karimullah] contaminated the evidence as part of his official duties as a police officer in the ALP."

There was further reason to doubt this identity as well. In the FOIA documents for Karimullah Abdul Karim, the facial image presents a young man in his early twenties. His date(s) of birth, which we know are somewhat unreliable, are listed as: "12/31/1992, 1/1/1993, 12/31/1993, and 1/1/1994," meaning this Karimullah was around twenty years old at the time of the murders. In the *Leavenworth* documentary, the third motorcycle rider, Haji Karimullah's son, tells his side of the story. His name is Mohibullah Aabid and he looks about twenty-five years old (circa 2018). Aabid says his father died from heart disease related to trauma from the July 2, 2012, incident. Father and son can't be so similar in age as Carney's ID would make them.

The information in a second FOIA release of records for "Ghamai" was equally revelatory. We know from the army's charge sheet that the man Lorance was in prison for murdering went by the name Ghamai Abdul Haq. The defense team shortened the name to "Ghamai," for ease, they said. Carney's Ghamai has the biometric identification (BID) B2JK9-B3R3. The data in this new FOIA record said this man's name was Ghamai Mohammad Nabi. The man's biometric link analysis chart linked him to six known Taliban bomb makers, but none of them were the family members of the men killed. More significantly, he hadn't even been killed on July 2, 2012; indeed, he'd been detained the four times between July 2012 and May 12, 2017.

As for the man Carney identified as Mohammad Rahim, FOIA docs listed him as Mohammad Rahim Sayed Wazir Toogh, a most wanted terrorist with a long list of criminal offenses and biometric captures. The Mohammad Rahim involved in the July 2, 2012, incident, shot in the arm in the Second Engagement, had his biometrics taken by Private Skelton that day. It is highly implausible that the army would pay solatia to him knowing he was a most wanted terrorist.

Now it is September 1, 2019, and I am sitting with John Maher at a restaurant along the Fox River, in a Chicago suburb called Geneva. Maher is of medium build, medium height, and when we meet for dinner he's wearing a pin-striped suit and a several-day beard. He is charismatic and smooth-talking, speaks fast. Enunciates every word. John Maher could have answered my questions on the telephone, which is how we had communicated since 2018; between us, there was, at this point, a record of 142+ emails. But after I presented him with this new batch of information that further contradicted what Lorance's defense team had warranted to the president of the United States, things reached a head.

"To answer your questions," Maher told me, "I need to see the whites of your eyes."

So I flew 2,015 miles from my home in Los Angeles to the Chicago suburb where he lives. Here I am. We are having dinner. Maher allowed me to record our interview.

"Where did Carney get the BIDs?" I ask. "How did he do his link analysis?" How does he explain that "Ghamai" is still alive? There is no straight answer from John Maher.

For ninety minutes, I go over all the FOIA documents I have with me. The claims the defense had made, and warranted to the president, are full of holes. What secret information was Bill Carney withholding that makes any of this make sense? I ask.

"Bill Carney can explain . . . everything," Maher promises.

Another month passes.

John Maher tells me cryptically, "There's a lot going on you don't know about, Annie."

On the morning of October 23, 2019, I woke up to a text message sent to me by Daniel Williams at 4:18 A.M.

It read: "Twist killed himself."

Twist? Impossible. How could this be? Twist, who was so happy. Twist, the father, the husband. The soldier who had moved on from the terrible things that had happened during the war. Twist, the law enforcement officer dedicated to the rule of law.

On the evening of October 22, James Twist shot himself in the head with his service revolver, in the master bedroom of the home where he lived with his wife and their three young children. He was taken by ambulance to the hospital, where he was declared dead the following morning. Because James Twist was an organ and tissue donor, his body was kept alive for several days while six of his major organs were harvested to save the lives of six critically ill individuals. In his horrific death, James Twist saved lives.

I was one of nearly a thousand people who attended the funeral service. There were hundreds of law enforcement and military reserve personnel packed into a church, many in formal dress uniform. Fourteen of James Twist's brothers from First Platoon were there. His extended tribe of family. Friends.

At the burial site, I observed his platoon mates throw clumps of earth into his grave. Twist's oldest son took his mom's hand.

"Where's Daddy?" he asked.

Thirteen days later, President Donald Trump pardoned Clint Lorance. After Lorance was released from the Disciplinary Barracks at Fort Leavenworth, John Maher and the defense team met the former first lieutenant at the prison gates and drove him in an SUV to a local hotel for a press conference. Clint Lorance blamed his incarceration on the "deep state." When asked about his fellow paratroopers of First Platoon, he said, "To be honest with you, I can't even remember most of their names."

For six months, Bill Carney ignored repeated requests for an interview. Then, in the middle of the pandemic, on April 24, 2020, he agreed to speak with me, from Helmand, Afghanistan, where he said he was. Maher and another member of the defense team were also on the telephone call.

By this time, Carney was aware that I was close to certain that the biometric information he presented to the defense team, and that the defense team presented to eight members of Congress and the president of the United States, was spurious. That it was not what it purported to be. I had written Carney a lengthy email with the facts laid out. By now, he knew I'd figured out that if he really had accessed the ABIS database, and found a so-called smoking gun, he could have been in violation of the Espionage Act.

In our call, Carney confirmed that he had not actually "punched the names" into the ABIS database after all, as John Maher said he had. Even more surprising, he told me that he wasn't even in Afghanistan at the time.

"I was not in the country," Carney conceded. "I was in my living room in Virginia."

So, where did he get the information on the bomb makers?

"When I left Afghanistan in July 2014," Carney said, "I burned a disc on all the files I had, for training. For rule-of-law training purposes. . . . I had about a hundred files on my laptop." An elaborate description of process ensued. Maher kept butting in, trying to make things sound less egregious than they actually were.

"I want you to know what was going through my head initially," Carney said in his own defense. "John said, 'Take a look at this, what do you think? . . .' He gave me the names [of the four men] . . . and I recognized the name Abdul Ahad right away."

Abdul Ahad was a "known bad guy" in southern Afghanistan, Carney insisted, which, if discussing the Abdul Ahad with the biometric identification (BID) B28JM-UUYZ, was true. It's just that the first fundamental of biometric identification is that a name means very little on its own.

In the *Leavenworth* documentary, the filmmakers tracked down and interviewed family members of the deceased motorcycle riders, including a man who identified himself as Abdul Ahad. I asked Carney if he honestly thought this was *the same* Abdul Ahad who was a most wanted terrorist marked for death by the U.S. government.

The same Abdul Ahad that Carney had based his questionable link analysis around.

"I can't say it's *not* him," Carney told me.

Then, in his own defense, Carney said, "The investigation [against Clint Lorance] was compromised from day one," as if the ends justified the means. "I really didn't see the charges against [him] as evidence-based."

I asked Carney how he did his link analysis.

"I filtered the information that I had . . . files of [Taliban] bomb makers, through an Excel spreadsheet," he said. "Incidents close to Zhari . . . I said to myself, let me see about any criminal activity . . . Like I said, I knew the name Abdul Ahad. That name had come up in [earlier] investigations. And I built out the links . . . the other names of the motorcycle riders, from there."

The information that Clint Lorance's defense team presented to the president of the United States was based entirely on the wrong Abdul Ahad, the wrong Ghamai, the wrong Haji Karimullah. Lorance had ordered civilians killed. He was convicted of these murders and jailed for war crimes. He was pardoned and freed by President Trump after eight members of Congress and the defense team presented information that stated Lorance had killed terrorists. And now, Carney confirmed, information used to free him was false. Two Afghan civilians, Haji Mohammed Aslam and his son, Ghamai Abdul Haq, were dead. A third member of the family, Haji Karimullah, had narrowly escaped death. In the name of justice for their already tried and convicted client, the defense team had falsely presented these men as being wanted terrorist bomb makers who'd killed Americans. But neither Carney nor Maher would take any responsibility for the harm done. Instead, they blamed biometrics.

"Biometrics is not an exact science," Bill Carney said.

But that is the whole point of the biometric identification. Fingerprints, iris scans, and DNA don't lie. Humans do.

"It's an imperfect science," Maher hedged. "Which is why . . .

before we went to the president, we reached out to certain members of Congress who have access to classified information. We did this through our Justice for Warriors Caucus. We counted on them telling us if our biometric information was wrong . . . We said, 'Find out and make sure it's right.'"

To that, Bill Carney added: "This is what happens when you have a compromised, contaminated investigation to begin with."

I asked the two men: "Do either of you feel regret or remorse that you portrayed murder victims as Taliban bomb makers?" Not just to the president, but to the countless Americans who donated money to a cause with a bogus biometric foundation.

John Maher asked me not to insult him, or Bill Carney.

Perhaps, to them, my question wasn't the point. If the defense team intended to sway the court of public opinion, that is just what they did.

The story of First Platoon is a ruinous tale of tragedy, hypocrisy, and loss. It did not start out this way. Each of the young men I interviewed for this book entered into the U.S. Army with enthusiasm and dedication. One hopes the living can move on, and the dead can rest in peace.

Except it is not over for the defense team. Its work continues, with a next quest involving Staff Sgt. Robert Bales: freeing him from what they say is a faulty war crimes charge. Bales, who sits in prison at Fort Leavenworth, and who pled guilty for killing sixteen Afghan villagers in cold blood.

"One of his victims, a child, he shot in the head from a few inches away," the army's lead prosecutor on the case, Lieutenant Colonel Joseph "Jay" Morse, told me.

John Maher and the defense team, with the support of United American Patriots, have built their appeal on "Fingerprint/DNA Impeachment Evidence." Data provided to them by Bill Carney, which claims that at least two of Bales's murder victims were Taliban bomb makers who left fingerprints and DNA on IED components. Their appeal is presently on file with the Supreme Court of the United States.

What will be the future of rule of law? America is becoming a society overwatched by a digital-genetic panopticon. What legal scholars describe as "a burgeoning 'National Surveillance State.'" There are more than 85 million ground-based surveillance cameras installed across America, the largest per capita share of surveillance devices globally, with more than one surveillance camera for every four people. The *Wall Street Journal* reports that by the end of 2021 there will likely be more than 1 billion surveillance cameras in operation around the world, 500+ million of them in China. Drones, aircraft, airships, and aerostats fly overhead recording video for later use. During a recent interview I did with Police Chief Bill Whalen from El Segundo, California, the Clearview AI demonstrator app he had on his iPhone was able to positively identify me less than one second after its facial-recognition software was pointed my way. The government's interest in biometric data has moved far beyond fingerprints, iris scans, facial images, and DNA to now include ear shape, voice, gait, odor, heartbeat, vein patterns, and more.

"Biometric cybersurveillance [is] increasingly used to justify the mass digital capture and analysis of unique physiological and behavioral traits of entire populations and subpopulations," says legal scholar and former Department of Justice attorney Margaret Hu. "The potential consequences [are] worrisome."

The argument that what is happening in China—that is, the mandatory data-banking of a whole population's biodata, including DNA—could never happen in America is an optimistic one. The pandemic of 2020 has resulted in enthusiasm for government-led contact-tracing programs in the U.S., opening the door for military-grade programs to data-bank biodata of Americans. Because disease lies at the center of this new threat, the reality that citizens' DNA cell samples are of interest to the government is no longer science fiction.

On April 10, 2020, the U.S. Department of Health and Human Services entered into a no-bid contract with Palantir Technologies to track the spread of the coronavirus. HHS is a cabinet-level, executive branch organization whose 2021 budget ($1.427 trillion) is more than twice that of the Defense Department ($705.4 billion). The goal of the HHS Protect Now program, said its spokesman, is to "bring disparate data sets together and provide better visibility to HHS on the spread of COVID." HHS confirmed the data that Palantir is now mining includes "diagnostic testing data; geographic testing data; [and] demographic statistics," meaning information about individual American citizens' health, location, family, and tribe. The initial HHS announcement said Palantir would be given access to 187 data sets. That number has since grown to 225. Unknowns abound: What data is going into the Palantir system, how is it shared, with whom, and for how long?

"Given how tight-lipped both HHS and Palantir have been over the program, we don't fully know," says Lauren Zabierek, executive director of the Cyber Project at Harvard Kennedy School's Belfer Center. Zabierek is a former U.S. Air Force officer who also served as a civilian analyst with the National Geospatial-Intelligence Agency, in three war zones, including in Kandahar in 2012. "I sincerely hope that HHS Protect Now will do nothing resembling finding and fixing certain entities," she says, using military nomenclature for locating and killing IED emplacers in the war zone. "I hope that [the data sets] will only be used to understand the spread of the virus in the aggregate."

As for Palantir, the privately held company seems to enjoy mythologizing the controversial nature of its work, its origins, and its founder, Peter Thiel. A large banner on the front page of the company website reads: "How a 'Deviant' Philosopher Built Palantir, a CIA-

Funded Data-Mining Juggernaut," an homage to the title from a *Forbes* magazine article from 2013.

For Americans, to adopt a "nothing to hide, nothing to fear" mindset ignores the reality that HHS has previously shared sensitive, personal information with federal law enforcement agencies. In 2018, HHS's Office of Refugee Resettlement allowed ICE to access confidential data files it had collected on migrant children, their family members, and potential sponsors after the federal government took the position that it needed this information to enforce the nation's immigration laws. To promote rule of law.

Ten weeks into the Protect Now program, a group of lawmakers wrote to HHS Secretary Alex Azar to express concern. "Unfortunately, HHS data has been misused before by federal law enforcement officials . . . We are concerned that, without any safeguards, data in HHS Protect [Now] could be used by other federal agencies in unexpected, unregulated, and potentially harmful ways." In a separate letter, members of the Congressional Hispanic Caucus were a little more blunt: "We have valid concerns on whether the existing surveillance framework Palantir has created to track and arrest immigrants will be supplemented by the troves of potentially personal health information contained within the HHS Protect [Now] platform."

It is the merging of disparate government databases into a giant monolith that has privacy experts concerned. "As nations enter into agreements to share biometric databases for military defense, foreign intelligence, and law enforcement purposes, the multinational cybersurveillance implications of biometric data collection and data analysis are likely to expand," warns Hu.

Shades of this merger are beginning to appear. Inaugurated in 2018, the Biometric Technology Center in West Virginia became the first joint biometric initiative between the Defense Department and the Justice Department to have its own physical building on U.S. soil. "With the opening of the BTC, the DoD and FBI will be able to work in collaboration to carry out operations and technical innovations to identify threatening or dangerous individuals," an army official said. The DoD's Defense Forensics and Biometrics Agency covers one-sixth of the 360,000-square-foot biometrics center; five-sixths of the center belongs to the FBI. And it is here where so many of the government's big-data systems are housed.

What does the ABIS machine actually look like? I wondered. This secretive behemoth, built of biometric data captured from millions of people in Iraq, Afghanistan, and elsewhere. Repeated requests to tour the facility, and to see ABIS with my own eyes, were denied. I asked Tom Bush about ABIS and its almost mythical stature.

"Lockheed built ABIS from old parts in the basement, and it still sits in the basement of CJIS," said Bush. "Lockheed built it because they built us," he said, meaning the FBI's database. "But DoD is a mess. So many holes. There's no standard biometrics collection like with CJIS. There should be a CJIS [a Criminal Justice Information Services Division] at DoD. There is not. This is what we argued back in '05, in '06. DoD has [what] is called a biometric gap. No one can get through the labyrinth."

Gaps and labyrinths. Vacuum-like spaces and mazes into which critical information can fall. A database that so few have access to, and even fewer understand, with information that can be improperly used to influence a decision made by the president of the United States. These are dangerous areas of operation in the complex geospatial terrain known as rule of law. As it stands now, in the criminal-justice system, individual humans are the backbone of the rule of law. In the systems of law enforcement, courts, and corrections, the humans still matter most. It is a human who commits the crime. A human who pulls the trigger or plants the IED. A human who finds the smoking gun. Automated machines do the grunt work, but humans solve the puzzle.

The process itself is like a labyrinth. Using science to solve crimes is thousands of years old, taking into account ancient autopsies and medieval toxicology reports. Using forensic biometrics to solve crimes has been part of the criminal justice system for 129 years, starting when Francisca Rojas left a bloody fingerprint on a doorway in 1892. The next twenty years of this century will almost certainly be transformative, taking leaps never before imagined.

A year 2020–2025 broad agency announcement, for advanced technology development programs for the army's Special Operations Command, provides a glimpse at what warfighting might be like in coming years. In this future operating environment, so-called hyper-enabled operators, or HEOs, will have systems on their bodies that allow for "persistent near-real-time collection" of biometric data to identify friend or foe. Fingerprint capture will be touchless, meaning the Pentagon's electronic systems will read people's ridges, whorls, and loops from "extended distances" without their awareness or consent. Soldiers will carry "rapid, portable, handheld DNA collection and processing [devices] for matching against authoritative databases." Translation: Dr. Selden's microwave-size, rapid DNA system will soon be the size of an iPhone, allowing for on-the-spot DNA checks.

One technology being pursued gave me pause: a "handheld, manpackable [machine] with holographic capabilities." The Pentagon aims to have its hyper-enabled operators carry a device that has "the ability to project images that are not real but seem real, and have the ability to develop personalized message campaigns for the image to project." In other words, three-dimensional deepfakes, to trick the enemy—in real time.

As it stands now, scientists have developed ways to identify what a person might look like from micrograms of captured DNA. Advances to build full-scale models of a person's probable identity from the gene up are ongoing. Add into the mix additional identity information—from ear shape to vein pattern, voice, gait, and skeletal frame—and this realistic-looking, soldier-projected hologram of a person is meant to be misinterpreted as the real person, the one with the true identity.

How much further will it go? What will society become? Will humans still recognize who—and what—the real villains really are?

ACKNOWLEDGMENTS

I first started thinking about military-grade biometrics while reporting my previous book, about the CIA. A U.S. Army Special Forces soldier turned CIA paramilitary operator told me an unusual story about biometric collection during a mission in Iraq. Insurgents in Mosul had taken over an airstrip formerly under U.S. control. This source was in charge of the Special Tactics Unit sent in to reclaim it. By the time the firefight for the airstrip was over, numerous enemy fighters were dead. The rest fled. The mission now shifted to biometric collection, the source explained, this being the way in which high-value targets are positively identified. For the operators, trying to collect fingerprint scans from half a dozen dead men in a war zone was time-consuming and dangerous. Iris scans were easier to get, the source told me, and the CIA's paramilitary operators had developed an unorthodox technique to accomplish this quickly. It involved two toothpicks. One operator would hold a dead man's face while a second operator used the toothpicks to get the iris into position. A third operator quickly scanned the eye using a handheld biometrics kit. "Everyone wants biometrics now," the source told me. "Collection time is dangerous. We [CIA operators] develop go-arounds because we can. Not so easy for young soldiers who must follow Big Army's ROEs." That got my attention, and the result is *First Platoon*.

That the Defense Department spends untold sums of U.S.

taxpayer dollars having young soldiers collect biometrics on entire populations and subpopulations in war zones was stunning to me then, and remains so now even after having written this book. Most Americans have no idea this is going on. Interestingly, a majority of the subject matter experts working in this field also remain largely in the dark. "It's the kind of thing we've *heard* about, but we don't have a lot of facts on," William Thompson, professor emeritus of criminology, law, and society at the University of California, Irvine told me in the spring of 2020. "The kind of thing my colleagues and I discuss over cocktails." This sentiment was echoed by former Department of Justice attorney Margaret Hu, presently professor of law and of international affairs at Pennsylvania State University. Hu is a subject matter expert on national security and cybersurveillance, and she coined the term "dataveillance," soon to enter the American zeitgeist, I surmise.

I am grateful to all the individuals who contributed to this book—every soldier, civil servant, federal agent, policymaker, professor, lawyer, law enforcement officer, historian, and others who let me interview them. Never have I experienced so many real-time plot twists while working on a book, some fortuitous, others tragic. In the words of the great Kurt Vonnegut, "So it goes." Vonnegut, who knew the horrors of war.

The commitment and generosity of those who work to preserve the record and make it available to the public never cease to amaze me. Thank you, Mrs. Erin Chidester, FOIA Division, U.S. Army Criminal Investigation Command, U.S. Army Crime Records Center; Dave Meyer and Linwood Matthews, Jr., Office of the Provost Marshal General; Alecia S. Bolling, Chief, U.S. Army Freedom of Information Act Office/FOIA Liaison; David Fort, National Archives and Records Administration. Thank you, Assistant Chief Stephen J. Hughes, New York City Police Department, who allowed me to spend a few days with the NYPD, interviewing veteran detectives and police officers so as to better understand rule of law. Thank you, Lauren Barten and Karen Burkhartzmeyer, for arranging for me to attend the International Symposium on Human Identification, where I was able to interview so many of the top forensic scientists and criminologists working in the field. Thank you, James Burke—one of my literary heroes—for talking to me about the history of science; Jay Morse, lead prosecutor, *United States v. SSG Bales*, for patiently answering so many questions about military law; Joe Meyers, chief inspector of the FDNY's explosives unit, for helping me to better understand explosives; Jesse Alpert, for including me on the trip to Fort Irwin; John Twist, for sharing with me the existential mystery of loss and of grief. Thank you, Jim Hornfischer, John Parsley, Linda Rosenberg, Steve Younger, Alan Rautbort, Tiffany Ward, and Matthew Snyder for all the work behind the scenes; Frank Morse, Andrew Feinstein, and Tim Moynihan for the early reads.

Thank you, Tom Soininen, for being what always feels like the world's most supportive and loving dad, and for teaching me to value books; Kathleen Silver and Rio Morse (my two mentors); Alice Soininen, Marion Wroldsen, and Keith Rogers (my three muses); Kirston Mann, Sabrina Weill, Michelle Fiordaliso, Nicole Lucas Haimes, and Annette Murphy (my fellow writers, always). Finally, and more important than anything else in this world: thank you, Kevin, Finley, and Jett. The only thing that makes me happier than reporting and writing books is the daily joy, wonder, and surprise I get from the three of you. You guys are my best friends.

INTERVIEWS AND WRITTEN CORRESPONDENCE

FIRST PLATOON AND SECOND PLATOON

(Rank During 2012 Deployment)

Spc. Brian M. Bynes

Cpt. Joseph Callahan IV

1st Lt. Grant M. Elliot

Spc. Todd A. Fitzgerald

Spc. Joseph "Doc" Fjeldheim

Spc. Brett M. Frace

Staff Sgt. Joshua D. Giambelluca

Spc. Alan Gladney

Pfc. Lucas Gray

Spc. Dallas L. J. Haggard

Spc. Brandon Krebs

1st Lt. Dominic Latino

1st Lt. Clint Lorance

Staff Sgt. Michael Herrmann (later McGuinness)

2nd Lt. Jared Meyer

1st Sgt. Joseph Morrissey

Pfc. Zachary Nelson

Spc. Anthony Reynoso

Spc. Cole Rivera

Pfc. James Skelton

Cpt. Patrick K. Swanson
Pfc. Zachary Thomas
Pfc. James Oliver Twist (promoted to Spc. in theater)
Pfc. Samuel I. Walley
Staff Sgt. Daniel Williams
Spc. David M. Zettel

Jack Ballantyne, PhD, forensic molecular geneticist, National Center for Forensic Science, University of Central Florida.

Patrick Biltgen, aerospace engineer, BAE Systems Intelligence Integration Directorate; subject matter expert on activity-based intelligence capabilities, National Geospatial-Intelligence Agency.

Donald Brown, attorney, Clint Lorance defense team; former U.S. Navy JAG officer; former Special Assistant U.S. Attorney.

James Burke, science historian, author, BBC broadcaster.

Thomas E. Bush III, assistant director, (ret.), FBI Criminal Justice Information Services Division; recipient of the FBI's Presidential Rank Award for Meritorious Service; board member, Biometric Technology Center, West Virginia.

William Carney, biometrics expert, Clint Lorance defense team; Command Counter Corruption Advisor (C3A), CENTCOM, TFSW/Camp Shorab, Helmand Province, Afghanistan.

Joseph DiZinno, director (ret.), FBI Laboratory, Quantico; former unit chief of DNA Analysis Unit II.

Maj. Bill Donahue (ret.), founder, United American Patriots, Innocent Warrior Project.

Mamoon Durrani, journalist, BBC Pashto.

Command Sergeant Major Robert Edwards, Garrison CSM, Fort Irwin National Training Center, Fort Irwin, California.

Andrew Guthrie Ferguson, subject matter expert on big data policing; professor of law, American University Washington College of Law.

Thomas Galati, assistant chief of intelligence, NYPD.

Kimberly Gin, Sacramento County coroner.

Kevin H, pattern of life expert; PGSS aerostat operator; Task Force Odin, Afghanistan.

John B. Hart, deputy chief, Intelligence Bureau, NYPD.

Matthew Hoffmann, special agent, U.S. Army Criminal Investigation Command.

Margaret Hu, professor of Law and of International Affairs, Pennsylvania State University; former Office of Special Counsel, Civil Rights Division, U.S. Department of Justice.

Stephen J. Hughes, assistant chief, commanding officer of Patrol Borough Manhattan South, NYPD.

Michael King, Crime Scene Unit deputy inspector, Joint Terrorism Task Force, NYPD/FBI.

John N. Maher, attorney, Clint Lorance defense team; former principal program manager, Justice Center in Parwan, Afghanistan; former criminal prosecutor, U.S. Army Judge Advocate General's Corps.

Ross McNutt, PhD, astronautical engineer; president and founder, Persistent Surveillance Systems; former U.S. Air Force officer; former Defense Department wide-area surveillance program developer.

Lisa Mertz, criminalist and DNA expert, Department of Forensic Biology, New York City Office of Chief Medical Examiner.

Joe Meyers, chief inspector, Explosives Unit, FDNY.

Lauren Mick, public affairs specialist, Office of the Special Inspector General for Afghanistan Reconstruction (SIGAR).

Kevin Mikolashek, attorney, Clint Lorance defense team; former U.S. Army officer; former criminal prosecutor, Department of Justice.

John Miller, deputy commissioner of intelligence and counterterrorism, NYPD.

Abdul Subhan Misbah, senior Afghan legal consultant to Departments of Defense, State, Justice (FBI), Justice Center in Parwan Prison, Afghanistan, 2009–2014; deputy director, Lawyers' Association of Afghanistan; current adviser, Supreme Court of Afghanistan (Stera Mahkama).

Lt. Col. Joseph "Jay" Morse (ret.), lead prosecutor, *United States v. SSG Bales*; former Staff Judge Advocate and Deputy SJA at the 101st Airborne Division and Fort Campbell, Kentucky.

James O'Neill, New York police commissioner.

P. Jonathon Phillips, PhD, electronic engineer, National Institute of Standards and Technology (NIST); research pioneer in the fields of computer vision, face recognition, biometrics, and forensics.

Steve Renteria, criminalist and DNA expert, Los Angeles County Sheriff's Department.

Rodolfo "Rudy" Rios, special agent, U.S. Army Criminal Investigation Command.

Muhammad Sajjad, PhD, scholar, Institute of Geology, University of the Punjab, Quaid-i-Azam Campus, Lahore, Pakistan.

Richard Selden, MD, PhD, geneticist and molecular microbiologist; founder, ANDE; inventor of Rapid DNA information system.

Paul J. Shannon, supervisory special agent (ret.), FBI; former director for law enforcement policy, White House Homeland Security Council.

Darby Stienmetz, criminologist, Forensic Science Division, Washoe County Sheriff's Office, Nevada.

William C. Thompson, PhD, subject matter expert on forensic science and human judgment; Professor Emeritus of Criminology, Law, and Society; Psychology and Social Behavior, and Law, University of California, Irvine.

Col. Ed Toy (ret.), chief of staff, Joint Task Force Paladin, Combined Joint Task Force 101; ISAF branch chief for counter-IED operations, Regional Command South, Afghanistan.

William Vickers, subject matter expert for biometrics and forensics, formerly advising the deputy undersecretary of defense for intelligence on biometrics capabilities; special assistant to the director and deputy director of the Biometrics Task Force, Department of Defense.

Bill Whalen, chief of police, El Segundo Police Department, California.

John D. Woodward, Jr., director (ret.), U.S. Department of Defense Biometrics Management Office; twenty-year CIA officer, Clandestine Service, Directorate of Science and Technology.

Lauren Zabierek, subject matter expert on activity-based intelligence; former U.S. Air Force officer; former National Geospatial-Intelligence Agency civilian analyst; executive director Cyber Project, Belfer Center, Harvard Kennedy School of Government.

Family of members of First Platoon: John Twist, Brooks Twist, Barbara Twist, Emalyn Twist, Emily Scott Robinson, Kelly Walley, Daniel Williams.

NOTES

Abbreviations Used in Notes

GOVERNMENT DOCUMENTS

Abdulhat Abdulhat (aka Abdul Ahad), BID: B28JM-UUYZ	U.S Army National Ground Intelligence Center. *BMAT Biometric Intelligence Analysis Report: Abdulhat Abdulhat (aka Abdul Ahad), BID: B28JM-UUYZ.* United States Army Intelligence and Security Command, March 18, 2013 (includes multiple declassification releases).
AR 15-6 SSG Bales Incident	Allen, Gen. John. R. "Report of Investigation IAW AR 15-6 Facts and Circumstances Surrounding Allegations of Shooting Afghan Civilians Outside Village Stability Platform Belambai," August 13, 2015.
CID, Lorance	U.S. Army, Criminal Investigation Command. *Summary of Investigative Activity, Basis for Investigation [. . .] Platoon Leader LT Clint LORANCE [et al].* Control No. 0254-2012-CID379-77688. (71 pages, not redacted, 07/04/2012–07/08/2013).
Commander's Report, Lorance	U.S. Army, Criminal Investigation Command. *Commander's Report of Disciplinary or Administrative Action: LORANCE, Clint Allen.* AR 190-45/AR 195-2. Crime

	Records Center, Quantico, VA, August 7, 2013. (535 pages includes sworn statements, interview worksheets, agent's activity summaries, photographs, etc.).
Ghamai Mohammad Nabi, BID: B2JK9-B3R3	International Security Assistance Force (ISAF) Joint Command (IJC). *IED Identification: Ghamai Mohammad Nabi, BID: B2JK9-B3R3.* Biometric Intelligence Program. United States Army Intelligence and Security Command (includes multiple declassification releases).
Karimullah Abdul Karim, BID: B2JK4-G7D7	International Security Assistance Force (ISAF) Joint Command (IJC). *IED Identification: Karimullah Abdul Karim, BID: B2JK4-G7D7. Biometric Intelligence Program.* United States Army Intelligence and Security Command (includes multiple declassification releases).
USA v. Lorance	*The United States v. First Lieutenant Clint Lorance.* Proceedings of a General Court-Martial. Fort Bragg, NC, April 25, 2013; July 30, 2013.

GOVERNMENT AGENCIES & AFFILIATES

ALSA Center	Air Land Sea Application Center
CENTCOM	U.S. Central Command
DoD	U.S. Department of Defense
DSB	Defense Science Board
GAO	U.S. Government Accountability Office
JCS	Joint Chiefs of Staff
NGA	National Geospatial-Intelligence Agency
NIJ	National Institute of Justice
NIST	National Institute of Standards and Technology
OIG	Office of Inspector General
SCOTUS	Supreme Court of the United States

3 "We captured": interview with Samuel Walley. All quotes from First Platoon and Second Platoon paratroopers are from my interviews conducted from 2018 to 2020, unless cited as being from a trial transcript or the CID investigation.

5 "Mud-brick buildings": interview with Patrick Swanson.

5 "Our minds are dark": Jonathan Addleton, *The Dust of Kandahar: A Diplomat Among Warriors in Afghanistan,* 12.

5 "Hobbesian": Lt. Col. Brian Petit, "The Fight for the Village: Southern Afghanistan 2010," *Military Review* (May–June 2011): 26.

5 ruled by a network: Carl Forsberg, "Counterinsurgency in Kandahar: Evaluating the 2010 Hamkari Campaign," *Afghanistan Report* 7, 23.

5 product was biometrics-enabled intelligence: Paul Moruza, "Intelligence Center Develops Biometrically Enabled Intelligence to Support Warfighter," Army.mil, January 9, 2013. The army also calls BEI an activity.

6 "biometrically enrolled by me": *USA v. Lorance,* 528.

7 different protocols: ALSA Center, *Biometrics: Multi-Service Tactics, Techniques, and Procedures for Tactical Employment of Biometrics in Support of Operations,* April 2014, 14–15; SEEK II Device Fingerprint Capture, training videos for Cross Match Technologies, author collection.

7 "within thirty minutes post-mortem": ALSA Center, *Biometrics: Multi-Service Tactics, Techniques, and Procedures for Tactical Employment of Biometrics in Support of Operations,* April 2014, 38.

7 Capturing facial images: ALSA Center, *Biometrics: Multi-Service Tactics, Techniques, and Procedures for Tactical Employment of Biometrics in Support of Operations,* April 2014, 13; U.S. Army Combined Arms Center, *Commander's Guide to Biometrics in Afghanistan: Observations, Insights, and Lessons,* Handbook No. 11–25, April 2011, 5, 40.

8 "mission specific BEWLs": ALSA Center, *Biometrics: Multi-Service Tactics, Techniques, and Procedures for Tactical Employment of Biometrics in Support of Operations,* April 2014, 3. Up to 40,000 biometric identities from the DoD's BEWL could fit on an individual SEEK in 2014.

9 “I thought we were in Afghanistan to kill”: interview with James Twist.

10 “At the start of the quarantine”: Michel Foucault, *Abnormal: Lectures at the Collège de France 1974–1975,* 45.

11 “All the information . . . uninterrupted power”: Michel Foucault, *Abnormal: Lectures at the Collège de France 1974–1975.* See also Michel Foucault, *Discipline and Punish: The Birth of the Prison,* 195–99.

13 “genetic panopticon”: SCOTUS, Syllabus, *Maryland v. King,* October Term, 2012. In Scalia, J., dissenting, 18, Scalia wrote, “Perhaps the construction of such a genetic panopticon is wise. But I doubt . . .”

14 set the record straight: Robert D. Olsen, Sr., “A Fingerprint Fable: The Will and William West Case,” *Identification News* 37, no. 11 (November 1987).

14 Bertillon measurements: NIJ, *Fingerprint Sourcebook,* 1-12. Photographs had been used for law enforcement purposes in Europe since around 1843, but it was Bertillon who standardized the technique for police bookings in 1888.

15 identical twins: “Name Index to Leavenworth Federal Penitentiary Inmate Case Files, 1895–1931,” National Archives at Kansas City, MO, Archives.gov.

16 emperor Qin Shi Huang: Bernhard Sonderegger and Martin Urs Peter, *The Fingerprint: 100 Years in the Service of the Swiss Confederation,* 13.

16 an ancient fingerprint: NIJ, *Fingerprint Sourcebook,* 1-12, 1-14, 5-5. Faulds was the first person to publish this important finding and its potential use as evidence. Later that same year, Sir W. J. Herschel wrote to *Nature* to point out that he had been using fingerprints in India for more than twenty years. “This prompted a battle of letters between Faulds and Herschel that would continue until 1917, when Herschel conceded that Faulds had been the first to suggest a forensic use for fingerprints.” Cited in “Fingerprints,” *Nature* 143 (February 25, 1939): 315. See also Henry Faulds, *A Manual of Practical Dactylography: A Work for the Use of Students of the Finger-Print Method of Identification,* The Police Review, 1923.

16 "finger-marks": Henry Faulds, "On the Skin-Furrows of the Hand," *Nature,* October 8, 1880. Faulds's letter to Darwin is dated February 16, 1880.

17 "Curved or whorled": Francis Galton, *Finger Prints,* 6.

17 Juan Vucetich: Digital images of the original documents that chronicle the facts of this story (which has also been turned into a legend) can be found at the National Institutes of Health exhibit "Visible Proofs, Forensic Views of the Body," NLM.NIH.gov.

18 the first person convicted on fingerprint evidence: Fingerprint card, Francisca Rojas (Individual dactiloscópica de Francisca Rojas), 1892, Dirección Museo Policial–Ministerio de Seguridad de la Provincia de Buenos Aires, Argentina, NIH exhibit "Visible Proofs, Forensic Views of the Body," NLM.NIH.gov.

19 Henry Classification System: NIJ, *Fingerprint Sourcebook,* 1-10, 5:8–9. The 422-page sourcebook is excellent for sorting fact from legend. Throughout, it pays tribute to obscure, original source material that is more generally overlooked (i.e., that a German professor named Dr. Johannes Purkinje had done similar pattern-type work decades before Henry and his team). Waquar A. Khan, "Indelible Imprints: The Genius from Khulna," *Daily Sun,* May 8, 2017.

19 America's fingerpint database: NIJ, *Fingerprint Sourcebook,* 6:1–3.

19 "Through this centralization of records": "The FBI and the American Gangster, 1924–1938," press conference, April 18, 1925, FBI.gov.

20 "FBI men reassuringly": "Sleuth School," *Time,* August 5, 1935; John F. Fox, Jr., "The Birth of the FBI's Technical Laboratory—1924 to 1935," FBI.gov.

20 Carl Voelker: Kenneth R. Moses et al., "Automated Fingerprint Identification System (AFIS), NIJ, *Fingerprint Sourcebook,* 6:3–5.

20 called a "scanner": "Fiftieth Anniversary of First Digital Image Marked," NIST, May 24, 2007. Kirsch's group invented the first image scanner in 1957. One of the first images scanned was that of Kirsch's three-month-old son, Walden, with a bit depth of one bit per

pixel. Considered an origin point in digital photography, in 2003, *LIFE* named this image one of the 100 photographs that changed the world.

21 "The deal was": interview with Tom Bush. Unless otherwise indicated, all quotes attributed to Tom Bush come from our interviews and email correspondence, 2018–2020.

22 Freedom of Information Act: "Your Right to Federal Records: Questions and Answers on the Freedom of Information Act and the Privacy Act, 1992," Electronic Privacy Information Center, EPIC.org.

23 David James Roberts: "Follow-up on the News; The 2 Identities of David Roberts," *New York Times,* September 25, 1988.

24 Woodrow Bledsoe: "In Memoriam, Woodrow Wilson Bledsoe," UTexas.edu; Michael Ballantyne, Robert S. Boyer, and Larry Hines, "Woody Bledsoe: His Life and Legacy," *AI Magazine* 17, no. 1 (Spring 1996): 7–20.

24 teaching machines to learn nuance: A. J. Goldstein, L. D. Harmon, and A. B. Lesk, "Identification of Human Faces," *Proceedings of the IEEE* 59, no. 5 (May 1971): 748–60.

25 installed a system: *FBI Law Enforcement Bulletin* 69, no. 6 (June 2000): 13.

25 finally created an algorithm: interview with Dr. P. Jonathon Phillips.

25 super-recognizers: P. Jonathon Phillips et al., "Face Recognition Accuracy of Forensic Examiners, Superrecognizers, and Face Recognition Algorithms," *PNAS* 114, no. 24 (June 12, 2018): 6171–176.

26 using the Daugman algorithms: John Daugman, "Major International Deployments of the Iris Recognition Algorithms: 1.5 Billion Persons."

29 Combined DNA Index System (CODIS): The FBI Laboratory's CODIS began as a pilot software project in 1990, serving fourteen state and local laboratories. In 1994, the DNA Identification Act formalized the FBI's authority to establish a National DNA Index System (NDIS) for law enforcement purposes. As of 2020, more than ninety law enforcement laboratories in over fifty countries use the CODIS software.

29 40 million: "Status of IDENT/IAFIS Integration," Report No. I-2002-003, December 7, 2001, USDOJ.org.

31 BODY PARTS: Of the initial 19,916 human remains, the New York City medical examiner identified 10,190 body parts, which totaled 58 percent of the victims. See also Jesse M. Raiten, MD, "Among Body Parts and Colleagues: Finding My Team in the Rubble on 9/11," *Anesthesiology* 129 (December 2018): 1186–88.

37 unbelievable match-hit: interview with Paul Shannon; interview with Tom Bush.

37 "I received a phone notification": interview with Paul Shannon.

37 Saudi national: "Unclassified Summary of Final Determination," Periodic Review Board, Detainee Name: Mohammed Mani Ahmad al-Qahtani, Detainee ISN 63, July 18, 2016.

37 Qahtani had been denied entry: Statement of Jose E. Melendez-Perez to the National Commission on Terrorist Attacks upon the United States, January 26, 2004. All the information to lawmakers in this section comes from this testimony.

39 Strategic warning assessment: Glenn J. Voelz, *The Rise of iWar: Identity, Information, and the Individualization of Modern Warfare,* U.S. Army War College, October 2015, 47.

39 Identity Dominance: interview with John Woodward; see also John D. Woodward, Jr., "Using Biometrics to Achieve Identity Dominance in the Global War on Terrorism," U.S. Army Combined Arms Center, September 2005. Author's note: Mohammed al-Qahtani was sent to the Guantánamo Bay detention camp in Cuba. In between enhanced interrogation sessions, which numbered at least fifty and twice brought him to the edge of death, Qahtani said he saw ghosts and was observed communicating with an imaginary bird. He and his fellow foot soldiers captured by the Pakistani Army, many of whom served as Osama bin Laden's sentinels, would become known as the Dirty Thirty. Their biometric profiles were the first of millions to come. See also Adam Goldman, "Saudi Suspected of Waiting to Aid 9/11 Hijackers Seeks to Leave Guantanamo," *Washington Post,* June 14, 2016.

39 Change happened fast: National Science and Technology Council, *Biometrics History,* March 31, 2006.

40 executive agent in charge: National Science and Technology Council, *Biometrics in Government Post-9/11: Advancing Science, Enhancing Operations,* September 11, 2008, 24.

41 "My goal was to get CIA": interview with Tom Bush; see also Owen Bowcott, "UK: FBI Wants Instant Access to British Identity Data," *The Guardian,* January 15, 2008.

42 using wet ink: interview with Paul Shannon.

42 buccal swab: Shaoni Bhattacharya, "Fast-Track DNA Tests Confirm Saddam's Identity," *New Scientist,* December 15, 2003.

43 U.S. claims proved illusory: "Final Report of the Commission on the Intelligence Capabilities of the United States Regarding Weapons of Mass Destruction," Report to the President of the United States, March 31, 2005. "A presidential commission concludes in March 2005 'not one bit' of prewar intelligence on Iraqi weapons of mass destruction panned out," wrote the Council on Foreign Relations in summarizing the Iraq war.

44 50,000 Iraqi detainees: Watson Institute for International and Public Affairs, Brown University, Costs of War Project, Watson.Brown.edu. This figure covers the first three years.

45 "When one was developed": Cheryl Benard et al., *The Battle Behind the Wire: U.S. Prisoner and Detainee Operations from World War II to Iraq,* Office of the Secretary of Defense, National Defense Research Institute, 2011.

45 went rogue: Seymour M. Hersh, "Torture at Abu Ghraib," *The New Yorker,* May 10, 2004. As of June 24, 2010, others were also involved; see the Taguba Report, the Schlesinger Report, and the Jones-Fay Army Report.

46 little-known organization: DoD, DSB, *Defense Science and Technology,* May 2002; DoD, Office of the Under Secretary of Defense for Acquisition, Technology, and Logistics, *Report of the Defense Science Board Task Force on Defense Biometrics,* March 2007.

46 "Are we winning": Donald Rumsfeld to Gen. Dick Myers, Paul Wolfowitz, Gen. Pete Pate, and Doug Feith, "Global War on Terrorism," memo, October 16, 2003.

46 "The [war] cannot be won": John D. Woodward, Jr., "Using Biometrics to Achieve Identity Dominance in the Global War on Terrorism," U.S. Army Combined Arms Center, September 2005, 32.

47 "Our military and intelligence concerns": DoD, Office of the Secretary, DSB, "Action: Notice of Advisory Committee Meeting," *Federal Register* 69, no. 26, Monday, February 9, 2004.

48 the Biometric Automated Toolset: Tom Dee, *NDIA: Disruptive Technologies: U.S. Army Intelligence Center of Excellence,* DoD, Biometrics, September 5, 2007. U.S. Army Intelligence Center of Excellence, "Biometrics Automated Toolset-Army (BAT-A) STRAP Increment 1."

49 Pentagon ordered 2,000: Biometrics Task Force 19, *Biometric Automated Toolset (BAT) and Handheld Interagency Identity Detection Equipment (HIIDE),* Overview for NIST XML & Mobile ID Workshop, September 2007.

49 battle for Fallujah: Benjamin Muller and John Measor, "Securitizing the Global Norm of Identity Biometrics and *Homo Sacer* in Fallujah," May 20, 2005; see also Ariana Dongus, "Galton's Utopia—Data Accumulation in Biometric Capitalism," *Spheres: Journal for Digital Cultures,* November 20, 2019.

50 "We'd conduct a quick": Paul J. Shannon, "Fingerprints and the War on Terror: An FBI Perspective," *Joint Force Quarterly* 43 (4th Quarter 2006): 82; interview with Paul Shannon. In between trips to Iraq, Shannon spent two years on the Homeland Security Council at the White House, where his focus was on interoperability. Collecting fingerprints in the field, during the height of the fighting, was brutal and bloody. His mom, a librarian in San Diego, California, asked him to please write home. "She didn't like email," Shannon told me, and he wanted to honor her wishes. "What I was dealing with was all classified, all tactical. So, I started writing poems and sending them home." His mom kept them all and compiled them into a book, *Songs of Iraq: A Year Long Deployment.* He submitted them to the Marine Corps Foundation and won an art award for best poetry written in a war zone. His mom accompanied him to the honorary banquet.

50 there was no place in the war theater: "TEDAC Marks 10-Year Anniversary," News, FBI.gov, December 12, 2013. TEDAC was created in late 2003.

50 ATF was in charge of: "FBI Names New Laboratory Director: DiZinnio Will Lead the Federal Bureau of Investigation Laboratory," FBI National Press Office, September 12, 2006; interview with Dr. Joseph DiZinno. Unless stated otherwise, DiZinno's quotes are from our interviews.

51 attorney general had assigned: Bureau of Alcohol, Tobacco, Firearms and Explosives (ATF), "ATF, Military, Expand Their Collaboration Against IEDs," Office of Public Affairs, February 27, 2007; *The FBI Laboratory 2006 Report,* An FBI Laboratory Publication, Federal Bureau of Investigation, Quantico, VA.

52 Rumsfeld told Congress: Committee on Armed Services, "Department of Defense Authorization for Appropriations for Fiscal Year 2007," Volume 4, Part 1, February–March 2006, Questions Submitted by Sen. John Warner to Secretary Rumsfeld, 90–91.

53 Pentagon purchased 1,648: Tom Dee, *NDIA: Disruptive Technologies: U.S. Army Intelligence Center of Excellence,* DoD, Biometrics, September 5, 2007, 1–34.

54 "Technologies that allow": Stew Magnuson, "Friend or Foe? Defense Department Under Pressure to Share Biometric Data," *National Defense Magazine,* January 1, 2009.

54 "the final layer of proof": "West Virginia on Cutting Edge of Latest Advances in Biometrics," *Capacity,* Fall 2006.

54 "We don't have a single belly button": Stew Magnuson, "Friend or Foe? Defense Department Under Pressure to Share Biometric Data," *National Defense Magazine,* January 1, 2009.

54 "We are still in the throes": Stew Magnuson, "Friend or Foe? Defense Department Under Pressure to Share Biometric Data," *National Defense Magazine,* January 1, 2009.

54 a congressional audit: GAO, *Defense Biometrics: DOD Can Better Conform to Standards and Share Biometric Information with Federal Agencies,* Report to Federal Requesters, GAO-11-276, March 2011, 1.

55 the Biometrics Knowledgebase System: Dan Caterinicchia, "DOD Opens Biometrics Site," *Federal Computer Week,* July 10, 2003.

55 "But why did DoD need": interview with Tom Bush.

56 "We found al-Qahtani": interview with Paul Shannon.

56 Lockheed Martin: John D. Woodward, Jr., "How Do You Know Friend from Foe?" *Homeland Science and Technology,* December 2004, 112.

56 $3.5 billion budget: GAO, *Defense Biometrics: DOD Can Better Conform to Standards and Share Biometric Information with Federal Agencies,* Report to Congressional Requesters, GAO-11-276, April 2012, Highlights, i.

56 1.5 million records: DoD, Biometrics Task Force, *Biometrics Task Force, Annual Report FY07,* 7.

57 2.7 million Iraqis: DoD, Biometrics Task Force, *Biometrics Task Force, Annual Report FY08,* 6.

57 "not operationally suitable": DoD, Director, Operational Test & Evaluation, *Automated Biometric Identification System (ABIS) Version 1.2: Initial Operational Test and Evaluation Report,* May 2015, i–iv.

57 "not survivable": DoD, Director, Operational Test & Evaluation, *Automated Biometric Identification System (ABIS) Version 1.2: Initial Operational Test and Evaluation Report,* May 2015, i.

57 instead of correcting: Tom Bush wrote a classified monograph on these problems, which I was unable to obtain through FOIA. Avoiding specifics, Bush confirmed with me that the DoD did not correct these problems.

57 more than 2 million people: Rod Nordland, "Afghanistan Has Big Plans for Biometric Data," *New York Times,* November 19, 2011. Initially, the system was to be called A-ABIS, for Afghanistan, but this was scrapped.

58 Terrorist Watch List 4: Rod Nordland, "Afghanistan Has Big Plans for Biometric Data," *New York Times,* November 19, 2011.

61 775,000 American service members: Lamothe, "How 775,000 U.S. Troops Fought in One War: Afghanistan Military Deployments by the Numbers," *Washington Post,* September 11, 2019.

62 "focus of insurgent activity": DoD, *Report on Progress Toward Security and Stability in Afghanistan,* December 2012, 23.

62 "You go in one way": interview with James Twist.

64 "It fit him perfectly": interview with John Twist.

64 first entry is ominous: John Twist sent me his son's journals and drawings after he committed suicide in 2019.

65 "Walking with all your gear": interview with Todd Fitzgerald.

66 raising the number: Danielle Kurtzleben, "CHART: How the U.S. Troop Levels in Afghanistan Have Changed Under Obama," National Public Radio, July 6, 2016.

66 "We cannot kill or capture": Robert Gates, "Future Military Strategy," National Defense University, September 29, 2008.

67 "Without ROL": U.S. Department of State and the Broadcasting Board of Governors Office of Inspector General, *Report of Inspection: Rule-of-Law Programs in Afghanistan,* Report No. ISP-I-08-09, January 2008, 1. In Afghanistan, the report found, rule of law was based on "different systems of dispute resolutions of [a] particular tribe or ethnicity, sometimes interwoven with Islamic law."

67 a series of monographs and reports: DoD, *Report on Progress Toward Security and Stability in Afghanistan,* December 2012, 23–24; Gretchen Peters, "Crime and Insurgency in the Tribal Areas of Afghanistan and Pakistan," Harmony Project, Combating Terrorism Center at West Point, October 15, 2010, 5.

67 "The infrastructure of the system": U.S. Department of State and the Broadcasting Board of Governors Office of Inspector General, *Report of Inspection: Rule-of-Law Programs in Afghanistan,* Report No. ISP-I-08-09, January 2008, 7.

67 "practice of *baad*": DoD, *Report on Progress Toward Security and Stability in Afghanistan,* December 2012, 112.

67 "Establishing the rule of law": Liana Sun Wyler and Kenneth Katzman, *Afghanistan: U.S. Rule of Law and Justice Sector Assistance,* R41484, Congressional Research Service, Report for Congress, November 9, 2010, summary page.

67 "[no] infrastructure system": Gretchen Peters, "Crime and Insurgency in the Tribal Areas of Afghanistan and Pakistan," Harmony Project, Combating Terrorism Center at West Point, October 15, 2010, 5.

68 "to centralize detention operations": Liana Sun Wyler and Kenneth Katzman, *Afghanistan: U.S. Rule of Law and Justice Sector Assistance,* R41484, Congressional Research Service, Report for Congress, November 9, 2010; see also Patrick J. Reinert and John F. Hussey, "The Military's Role in Rule of Law Development," *Joint Force Quarterly* 77 (2nd Quarter 2015): 120–127.

68 Rule of Law Field Force-Afghanistan: Mark Martins, "NATO Stands Up Rule of Law Field Support Mission in Afghanistan," *Lawfare Blog,* July 6, 2011.

68 Task Force Biometrics: CENTCOM, "Conference Maps the Way Ahead for Biometrics in Afghanistan," October 15, 2010.

69 Mohammad Anwar Moniri: CENTCOM, "Conference Maps the Way Ahead for Biometrics in Afghanistan," October 15, 2010.

70 80 percent: CENTCOM, "Conference Maps the Way Ahead for Biometrics in Afghanistan," October 15, 2010.

70 Fourth Amendment violation: interview with Margaret Hu.

72 falsely fingered: A 2007 documentary film about Dilawar's torture and death, *Taxi to the Dark Side,* won an Academy Award and earned the jail at Parwan the name Afghanistan's Guantánamo Bay.

73 "Families are allowed": Patrick J. Reinert and John F. Hussey, "The Military's Role in Rule of Law Development," *Joint Force Quarterly* 77 (2nd Quarter 2015): 120–127.

73 "part of the overarching goals": Ladonna S. Davis, "Team in Afghanistan Named Best Project Development Team in U.S. Army Corps of Engineers," U.S. Army Corps of Engineers, August 1, 2011.

73 the courts component: "Rule of Law Programs in Afghanistan," Department of State, Fact Sheet, May 4, 2012; see also the Justice Sector Support Program (JSSP), which utilizes fifty-eight U.S. justice advisers and 110 Afghan legal advisers to train and build capacity for Afghan officials within the Ministry of Justice, Attorney General's

Office (AGO) Ministry of Interior (MOI), Supreme Court, and Ministry of Women's Affairs.

74 the workers got physical: "NATO Apologises for Afghan Koran 'Burning,'" BBC Asia, February 21, 2012.

74 "We attacked them": Sangar Rahimi and Alissa J. Rubin, "Koran Burning in NATO Error Incites Afghans," *New York Times,* February 21, 2012.

74 "When the Americans insult us": David A. Krooks, Lucy A. Whalley, and H. Garth Anderson, "Contingency Bases and the Problem of Sociocultural Context," U.S. Army Corps of Engineers, ERDC/CERL TN-12-2, July 2012, 2.

75 "There are some crimes": Alissa J. Rubin, "Chain of Avoidable Errors Cited in Koran Burning," *New York Times,* March 2, 2012.

75 Karzai threatened: R. Chuck Mason, "Status of Forces Agreement (SOFA): What Is It, and How Has It Been Utilized?" Congressional Research Service, March 15, 2012, 1.

77 violence and anarchy: Murray Brewster, "'Au Revoir, Zangabad,' Canadian Army Hands Over Afghan Village to U.S." Canadian Press, June 2011.

78 "Justice and Liberty are held hostage": Geoff Demarest, *Winning Insurgent War: Back to Basics,* ii. The book's tone can be interpreted numerous ways. I found the dedication telling: "To Chuck's arm, which he nobly left on the other side of the world."

79 "topography": Geoff Demarest, *Winning Insurgent War: Back to Basics.* This quote is from "About the Cover Illustration" of the book, and was written by Charles A. Martinson III, who is, incidentally, the author of "The Topology of Quantum Timespace: A Theory of Everything."

79 "Anonymity is": Geoff Demarest, *Winning Insurgent War: Back to Basics,* 2.

79 "If at all possible": Geoff Demarest, *Winning Insurgent War: Back to Basics,* 4.

79 "A peaceful liberal social contract": Geoff Demarest, *Winning Insurgent War: Back to Basics,* 5.

80 the pressure-plate improvised explosive device: International Campaign to Ban Landmines, "Landmine & Cluster Munition Monitor

Fact Sheet," November 2011; JCS, "Joint Publication 3-42: *Joint Explosive Ordnance Disposal*," September 9, 2016.

81 "merely a procedure": JCS, "Joint Publication 3-42: *Joint Explosive Ordnance Disposal*," September 9, 2016, 2-1. The army's best countermeasures in 2012 were amputation mitigation. Because PPIED blasts were amputating limbs on the spot, the army designed knee-high socks to make these battlefield amputations more succinct. NCOs were ordered to make sure soldiers wore "dick plates," Kevlar plates fitted for the genitals, to keep soldiers from losing their reproductive organs to war.

82 just a wooden pole: AR 15-6 SSG Bales Incident, Photo 10: Southerly view exiting the gate ECP, 435.

82 playing tricks: AR 15-6 SSG Bales Incident, 207, 213.

83 "unable to stand up": AR 15-6 SSG Bales Incident, Exhibit A, "VSO-ALP Methodology."

83 watching what happened to Asbury: Brendan Vaughan, "Robert Bales Speaks: Confessions of America's Most Notorious War Criminal," *GQ*, October 21, 2015.

84 "An American shot them": AR 15-6 SSG Bales Incident; interview with Col. Jay Morse, lead U.S. Army JAG prosecutor, *Bales v. USA*.

84 to alert medical personnel: AR 15-6 SSG Bales Incident, 362.

85 broken and awaiting parts: AR 15-6 SSG Bales Incident, 435.

85 Developed for the war in Afghanistan: "Summary Report of DoD Lighter-Than-Air-Vehicles," Office of the Assistant Secretary of Defense for Research and Engineering, Rapid Reaction Technology Office, October 2015, i–iii.

86 heard gunfire: AR 15-6 SSG Bales Incident, CID Exhibit 58, March 16, 2012. "Looking for 1x US soldier," the operator noted.

86 "The person [we are looking for]": AR 15-6 SSG Bales Incident, CID Exhibit 187, March 18, 2012, sworn statement, Navmar employee.

86 "[We began] placing chest seals": AR 15-6 SSG Bales Incident, 361–62.

87 Pave Hawk: AR 15-6 SSG Bales Incident, 40, 362.

87 "But the [younger] girl": interview with Mamoon Durrani. There are reports of this girl speaking via video chat to CID investigators. I consider Durrani the authority. With director Lela Ahmadzai, Durrani filmed and produced a short film called *Silent Night: The Kandahar Massacre,* which features the testimony of the children who witnessed the massacre, and who survived, as well as other relatives.

88 covered in blood: AR 15-6 SSG Bales Incident, CID Exhibit 45, 184–85. Detailed list of twenty-one items Bales was wearing or carrying, as well as a photograph of a bearded Bales, without eyeglasses, taken after his return to the VSP.

89 army paperwork: AR 15-6 SSG Bales Incident, 23, Appendix, 097.

89 His assignments included: AR 15-6 SSG Bales Incident, 28.

89 soaked in his victims' blood: AR 15-6 SSG Bales Incident, CID Exhibit 123, March 13, 2012.

89 videos of war dead: AR 15-6 SSG Bales Incident. See also Damian Cave, "250 Are Killed in Major Iraq Battle," *New York Times,* January 29, 2007, and Melisa Goh, "Former Captain: Afghan Shooting Suspect Showed 'Valorous Conduct' in Battle," National Public Radio, March 19, 2012.

90 Criminal Investigation Command, known as CID: to note, it's CID, not CIC, because it used to be the Criminal Investigation Division.

90 "Unfortunately, we are the army soldiers": Matthew Hoffman spoke to me briefly during a telephone call; the army prohibited him from speaking with me further.

92 "One of the little girls": interview with Mamoon Durrani.

92 Colonel Wood observed: AR 15-6 SSG Bales Incident, 348–350.

93 strange kind of symmetry: AR 15-6 SSG Bales Incident, CID Exhibit, March 13, 2012, Agent's Investigation Report; CID Exhibit, March 31, 2012, sworn statement by the STOF-S Operations Center Night Shift NCOIC.

93 body armor: AR 15-6 SSG Bales Incident, CID Exhibit 263, 2. The group was accompanied by BG Raziq's deputy, who told CID investigators what happened. "During a shura inside the Naja Bien mosque," Taliban opened fire on the delegation, killing one of General

Karimi's soldiers and wounding two National Defense Service investigators.

93 official Afghan Uniformed Police report: AR 15-6 SSG Bales Incident.

94 "Money is my most": U.S. Army, "Commander's Guide to Money as a Weapons System: Tactics, Techniques, and Procedures," Handbook No. 09-27, U.S. Army Combined Arms Center, April 2009, 1.

95 sipped tea: "Tragic Incident in Panjway District of Kandahar," YouTube, March 17, 2012, https://www.youtube.com/watch?v=SbueAcJYMZo, accessed June 1, 2019.

95 army's incorrect count: Rod Nordland, "3 NATO Soldiers Killed by Afghan Security Officers," *New York Times,* March 26, 2012. "Early in the day, an Afghan police official in Kandahar Province, where the killings took place, said the 17th victim could be accounted for because a pregnant woman was among the dead. But he later retracted that assertion, and American military officials restated that their investigation showed evidence for 17 murder charges."

95 Afghan version of events: AR 15-6 SSG Bales Incident, Special Operations Task Force, South, Commander's sworn statement, March 31, 2012.

95 solatia payments: AR 15-6 SSG Bales Incident, 348–350, 354.

96 "Habibullah needed": interview with Mamoon Durrani, photos provided by Durrani.

96 "The Taliban has started boiling people": interview with Mamoon Durrani, photos provided by Durrani.

97 "The whole area down there": interview with Grant Elliot.

98 ancient karez: Abobakar Himat and Selim Dogan, "Ancient Karez System in Afghanistan: The Perspective of Construction and Maintenance," *Academic Platform Journal of Engineering and Science,* March 29, 2019.

101 "I conduct BDA": Skelton, *USA v. Lorance,* 548.

101 no official training program: Phil Sussman, "COIST Staffs Play Crucial Role on Today's Complex Battlefield," U.S. Army press release, June 19, 2009.

101 storyboards: *USA v. Lorance,* 554. "I would do the storyboard after each mission involving 1 Charlie. I would do the storyboards, and I would do COIST intel analyst comments," said Skelton.

101 Soldiers chosen for the job: *USA v. Lorance,* 7.

102 overweight: Skelton's weight was brought up by numerous soldiers, which is why it is included here. In interviews with two officers, neither could account for how or why this was not addressed, saying only that every soldier is required to pass PT requirements at Fort Bragg.

103 "We were supposed to walk": David H. Petraeus, "COMISAF's Counterinsurgency Guidance," Headquarters, International Security Assistance Force/United States Forces–Afghanistan, August 1, 2010. 3.

113 "You could almost forget": interview with Anthony Reynoso.

115 "The government in Kabul": Peter Bergen, "The Crossroads: Can We Win in Afghanistan?" *The New Republic,* May 3, 2011.

116 "These people are farmers": interview with Dallas Haggard.

116 "'Hey, we want eye scans'": interview with Michael McGuinness (formerly Herrmann).

117 "We'd watch this activity": interview with Jared Meyer.

118 "We're going to find ourselves": Patrick Biltgen and Stephen Ryan, *Activity-Based Intelligence: Principles and Applications,* 2.

118 In a single year: Christopher Drew, "Military Is Awash in Data from Drones," *New York Times,* January 10, 2010.

119 "Well, we built Palantir": "CNBC Exclusive, CNBC Transcript: Palantir Technologies Co-founder & CEO Alex Karp Joins CNBC's Josh Lipton for a Rare Interview Airing Today," CNBC News Releases, February 28, 2018.

119 countless future wars: Barack Obama, "National Security Strategy," The White House, May 2010, 37. "America's commitment to the rule of law is fundamental to our efforts to build an international order that is capable of confronting the emerging challenges of the 21st century."

120 "Palantir went from": "CNBC Exclusive, CNBC Transcript: Palantir Technologies Co-founder & CEO Alex Karp Joins CNBC'S Josh

Lipton for a Rare Interview Airing Today," CNBC News Releases, February 28, 2018.

122 "Zhari is probably": Canadian Press, "U.S. Troops Take Root in Former Canadian Base," CTV News, August 22, 2010; Rajiv Chandrasekaran, *Little America: The War Within the War for Afghanistan,* 281.

122 ten-mile-long concrete wall: Rajiv Chandrasekaran, "In Afghanistan's South, Signs of Progress in Three Districts Signal a Shift," *Washington Post,* April 16, 2011; Ben Gilbert, "Afghanistan War: Bulldozing Through Kandahar," Public Radio International, November 10, 2010.

122 "[Come] to the claims center": "Working Through the Claims Process in Zharay," *The Heartbeat* 8, no. 4 (March 2011): 11–12.

122 thirty to forty barriers: "Flushing Out Taliban," *The Heartbeat* 8, no. 4 (March 2011): 18.

124 "Because of the high-powered equipment": interview with Kevin H.

125 underbelly of the aerostat: "Summary Report of DoD Funded Lighter-Than-Air-Vehicles," Office of the Assistant Secretary of Defense for Research and Engineering, Rapid Reaction Technology Office, October 2015, 21–27. PGSS is a U.S. Navy–owned aerostat. Sometimes the camera suite was the MX-20, with variant technologies.

127 aerostat operations at Siah Choy: interview with Grant Elliot.

128 Navmar Applied Sciences: With just 200 reported employees, Navmar secured $1.23 billion in defense contracts between 2007 and 2012, sending PGSS operators like Kevin H to forward-deployed combat outposts across Afghanistan, like the one at Siah Choy.

128 Navmar's ArcticShark: Sophia Chen, "This Drone Once Fought Wars. Now It's Fighting Climate Change," *Wired,* May 1, 2017.

129 an amalgamation: Kelly Kemp, "Clapper Inducted into the NGA Hall of Fame," *Pathfinder,* May 2008. In July 2004, Clapper published a working doctrine for GEOINT—the first ever Geospatial Intelligence Basic Doctrine.

131 highly prized IMINT: Office of the NGA Historian, "The Advent of the National Geospatial-Intelligence Agency," September 2011, 5–8.

131 U-2 spy plane's ability: "Agency U-2 Pilots: Hervey Stockman," From the Vault, CIA.gov; Annie Jacobsen, *Area 51: An Uncensored History of America's Top Secret Military Base.*

132 an entire year: Stew Magnuson, "Military 'Swimming in Sensors and Drowning in Data,'" *National Defense Magazine,* January 1, 2010.

132 an unlikely source: interview with Patrick Biltgen; Patrick Biltgen and Stephen Ryan, *Activity-Based Intelligence: Principles and Applications,* 24.

132 "card cheats": Patrick Biltgen and Stephen Ryan, *Activity-Based Intelligence: Principles and Applications,* 24. The classified white paper is called "Surveillance Employment Strategies for Irregular Warfare."

132 "[NGA] originally called this": interview with Patrick Biltgen; Patrick Biltgen and Stephen Ryan, *Activity-Based Intelligence: Principles and Applications,* 24.

133 defined by an act of Congress: interview with Patrick Biltgen; Patrick Biltgen and Stephen Ryan, *Activity-Based Intelligence: Principles and Applications,* 25.

133 wide-area surveillance system ARGUS-IS: interview with Patrick Biltgen.

134 S2 would monitor the situation: interview with Grant Elliot.

137 "The moment there is a target of opportunity": interview with Kevin H.

138 Lt. Grant Elliot: interview with Grant Elliot.

141 "the discovery of true identities": JCS, "Joint Publication 2-0: *Joint Intelligence,*" October 22, 2013, B-9. Note: Within the text, this concept can be noted as in effect since June 22, 2007. See page iii, "Summary of Changes, Revision of Joint Publication 2-0, Dated 22 June 2007." "Added a description of 'identity intelligence' and grouped in under production categories." A Top Secret document entitled "Identity Intelligence" was disclosed by Edward Snowden and published in the *New York Times* on May 31, 2014. See also Stephen Aftergood, "Identity Intelligence and Special Operations," FAS.org, July 30, 2014.

142 "connecting individuals": JCS, "Joint Publication 3-05: *Special Operations,*" July 16, 2014, IV-2, 4 (I2).

142 Abdul Ahad first came: *Abdulhat Abdulhat (aka Abdul Ahad), BID: B28JM-UUYZ*. All information in this chapter about Ahad comes from this report, which contains numerous intelligence products.

146 behind a maroon door: Maj. Carol McClelland, "Improved Training Helps Forensics Team Prepare for Afghanistan Deployment," U.S. Army, March 24, 2011.

147 they were "engaged": *Abdulhat Abdulhat (aka Abdul Ahad), BID: B28JM-UUYZ*. There are numerous incongruities in this record, likely a product of typographical error. They include the location of initial March detainment (Zhari versus Maiwand); the date of the RCIED recovery (November 17 versus November 26).

147 new forensics laboratory: *Abdulhat Abdulhat (aka Abdul Ahad), BID: B28JM-UUYZ*. Originally called CEXC-K, the facility was later renamed ACME-K.

147 "You have to remember": interview with Col. Ed Toy (ret.); Ed Toy, *Pressure Plate: A Perspective on Counter IED Operations in Southern Afghanistan 2008–2009*, 92.

148 "It may be how the bomb maker": interview with Col. Ed Toy (ret.); Ed Toy, *Pressure Plate: A Perspective on Counter IED Operations in Southern Afghanistan 2008–2009*, 94.

148 mobile DNA lab: Ed Drohan, "CSI Afghanistan: Forensic Experts Help Turn Bomb Maker into Convict," Defense Visual Information Distribution Service, December 10, 2013.

150 "DNA is 99.9999999999": interview with Dr. Richard Selden.

150 "The subject is assessed": *Abdulhat Abdulhat (aka Abdul Ahad), BID: B28JM-UUYZ* (Identity Name KAF-13-0248 [IED-ZHA-059610-13]).

151 "an almost industrial-scale": Gretchen Gavett, "What Is the Secretive U.S. 'Kill/Capture' Campaign?" PBS *Frontline*, June 17, 2011.

153 "All the IEDs were all marked": interview with Kevin H.

154 "Colonel Mennes's attitude": interview with Patrick Swanson.

154–55 "The intent that we set out": interview with Jared Meyer.

156 Dickhut was dead: interview with Kevin H.

157 blast wave: Craig Freudenrich, PhD, "How IEDs Work," HowStuffWorks: Science, December 10, 2008.

157 Mennes didn't want the soldiers distracted: interviews with Jared Meyer, Patrick Swanson, and Daniel Williams.

158 Olivas's death cast a long shadow: This chapter is based on interviews with the individuals involved, as quoted.

171 he was excited: CID, Lorance, 7/15/2012. In July 2012, Katrina Lucas was a second lieutenant. During the trial, when she testified, she was a first lieutenant.

171 Lucas later told a military judge: *USA v. Lorance,* 770–73.

172 a Ranger tab: *USA v. Lorance,* 773, 782.

172 Lucas knew this was a real issue: *USA v. Lorance,* 786.

172 "only about five weeks": *USA v. Lorance,* 784–85.

172 "Lorance was enthusiastic": CID, Lorance, 7/8/2012.

173 "Our AO [area of operations] was a little unique": *USA v. Lorance,* 777–78. For more on GRINTSUM, see U.S. Army Field Manual 34-8-2, Appendix E.

173 "He sent an email": *USA v. Lorance,* 774. Because a BOLO does not have the same degree of classification that a BEWL has, Afghan National Army soldiers and Afghan Local Police officers have access to the BOLO.

173 "high-payoff targeting list": *USA v. Lorance,* 774.

174 "Red motorcycles would": *USA v. Lorance,* 775, 797.

175 He'd been on the Joint Prioritized Effects List: *Abdulhat Abdulhat (aka Abdul Ahad), BID: B28JM-UUYZ* (note: JPEL packet code IS3824).

175 "He is part of a larger [IED] network": *Abdulhat Abdulhat (aka Abdul Ahad), BID: B28JM-UUYZ* (note: JPEL packet code IS3824).

176 crushing in its wickedness: interviews with the individuals involved, as quoted. "We were always trying to get them to tell us where IEDs were. But they didn't do that very much. Their lives were very much in danger as ours were, from the Taliban," per Fitzgerald.

178 "attempt[ed] to relax": Don Brown, *Travesty of Justice: The Shocking Prosecution of Lt. Clint Lorance,* 81.

178 back-to-back suicide bombings: Bashir Ahmad Nadem, "21 Killed, 50 Injured in Twin Suicide Blasts," *Pajhwok Afghan News,* June 6, 2012; Bill Roggio, "Taliban Kill 21 Afghans in Double Suicide Attack," *Long War Journal,* June 6, 2012.

178 killed twenty-three people: Alissa J. Rubin and Taimoor Shah, "Afghanistan Faces Deadliest Day for Civilians This Year in Multiple Attacks," *New York Times,* June 6, 2012.

180 "He [had] a little tiny, tiny kid": *USA v. Lorance,* 160; interviews with James Twist. (Note: For reader clarity, I removed from the transcript words that are doubled/repeated, as well as "like," "um," and "well.")

181 Afghan interpreter: CID, Lorance, 7/7/2012 (Fawad Elvis Omari). In this interview, Omari appears to be merging two episodes at the entry-control point: this one from June 30 and the one from the morning of July 2. The first refers to C-wire; the second is about soldiers shooting into the village.

181 moved the army's C-wire: interview with James Twist.

181 "Move the C-wire": *USA v. Lorance,* 162.

182 "We can make a deal": This dialogue comes from the trial transcript, *USA v. Lorance,* 162.

182 "You can tell us where": *USA v. Lorance,* 162–63.

182 "I know of no IEDs": *USA v. Lorance,* 182.

182 "He didn't really want to say": *USA v. Lorance,* 163, 256.

183 "He looked out": interview with Dallas Haggard. The "Nazi-style" comment was also discussed at length during pretrial arguments.

183 "to interdict": Swanson, *USA v. Lorance,* 704.

183 with the Minehound: CID, Lorance, 8/8/2012.

184 "a large metal object": CID, Lorance, 8/8/2012.

184 Ali told investigators: CID, Lorance, 7/16/2012. Ali gave his testimony to Special Agent Kevin Mitchell through an interpreter named Lal "Jack" Mohammad. Also of note: "Lieutenant Lorance asked what was the standard makeup of an IED in the area." Sergeant Peters told him it was seven or eight liters of explosive per IED.

Lorance reported to headquarters that he'd just found an IED with thirty pounds of explosives and a pressure plate.

185 return fire with the Carl Gustaf: interview with Michael McGuinness (formerly Herrmann). The MK-19 is from CID.

186 "why we were shooting at them": *USA v. Lorance*, 256; Prosecution Exhibit 17 photos.

186 "Recon by fire": *USA v. Lorance*, 153, 291; CID, Lorance, 7/14/2012.

186 never fired a shot: interview with David Zettel.

187 "One was a civilian": *USA v. Lorance,* 259.

187 "'You know, they're kids'": *USA v. Lorance,* 264–65.

188 "I saw our SDMs": Williams, *USA v. Lorance,* 295.

189 "I woke him up": *USA v. Lorance,* 297–98.

189 "[Lorance] had told [Zettel and Rush]": *USA v. Lorance,* 297.

189 "Lieutenant Lorance came [back] in": *USA v. Lorance,* 298.

189 "He told me the reason": *USA v. Lorance,* 300.

189 "Lorance said that he was in love": *USA v. Lorance,* 299–300.

190 (OPSEC) meeting: Swanson, CID, Lorance, 8/13/2012; *USA v. Lorance*, 702–6.

190 "Proportionality is using": *USA v. Lorance*, 705–6. "The platoon received fire from an AK-47 and responded with a MK-19."

191 "I made a point": *USA v. Lorance*, 706.

192 Lorance grew angry: interview with Todd Fitzgerald.

193 "Get Payback for Hanes": Gray, CID, Lorance, 7/7/2012.

193 "We had pre-deployment classes": interview with James Twist.

193 "Somebody objected": *USA v. Lorance,* 449.

193 "any two-wheeled motorcycle": *USA v. Lorance,* 302.

193 "We all knew something was wrong": *USA v. Lorance,* 449.

194 "'Shock-and-awe'": *USA v. Lorance,* 527.

194 "spectacular display": The idea comes from a 1996 Pentagon briefing document written by defense analysts Harlan Ullman and James

Wade. In March 2003, during a briefing in Qatar, Gen. Tommy Franks said of the Iraq invasion: "This will be a campaign unlike any other in history. A campaign characterized by shock, by surprise, by flexibility, but the employment of precise munitions on a scale never before seen, and by the application of overwhelming force."

194 Lorance ordered the two gun trucks: *USA v. Lorance,* 317. Also see statements of Spc. Anthony Reynoso, Pvt. David Shilo, and Spc. Brett Frace.

194 came in with a request: U.S. District Court, District of Kansas, *Clint A. Lorance v. Commandant,* Declaration of Kevin H. Case No. 18-3297-JWL, May 19, 2019.

195 "received ICOM chatter": JOCWatch entry 0184, July 2, 2012 (SIGACT), author collection.

195 already exited the strongpoint: interview with Dallas Haggard.

196 "They were known to me": *USA v. Lorance,* 527–28.

196 "I didn't hear the conversation": *USA v. Lorance,* 308–9.

197 "between 30, 35, 40": *USA v. Lorance,* 537, 571. Skelton gave at least five different statements to criminal investigators, reporting the speeds of the vehicle being as low as 10 miles per hour. Noted at trial was that Skelton was a traffic cop for two and a half years, with the Southern Pines, NC, Police Department and therefore had an understanding of estimating speed (571–78).

197 "The first thought I had": *USA v. Lorance,* 537.

197 "I fired two rounds, missing both": *USA v. Lorance,* 530. Ten days prior, Skelton dropped his rifle and had neglected to resight it, he told the military court.

197 "[I] had clearance over the grape row": *USA v. Lorance,* 451.

198 "I witnessed the motorcycle": *USA v. Lorance,* 452.

198 *What is going on?*: interview with Todd Fitzgerald.

198 "Lieutenant Lorance ordered us to open fire": *USA v. Lorance,* 451.

199 "He notified us that there was three PAX": *USA v. Lorance,* 731.

199 "Not necessarily like they were scared": *USA v. Lorance,* 736.

200 "They began pointing at our formation": *USA v. Lorance,* 317.

200 "I recognized the older gentleman": *USA v. Lorance,* 318.

200 "Lieutenant Lorance contacted the gun truck": *USA v. Lorance,* 668–69.

200 "that's when I was given the order": *USA v. Lorance,* 738.

201 "The second [man] went to go look": *USA v. Lorance,* 739.

201 Fitzgerald watched the third man: *USA v. Lorance,* 452–53.

201 tree line to the north: *USA v. Lorance,* 452.

203 "A small moment after": *USA v. Lorance,* 367–69.

204 "I asked First Lieutenant": *USA v. Lorance,* 370.

204 McNair later testified: CID, Lorance, 7/19/2012, 09:55. By trial, McNair was promoted to captain. "Photographs were taken of the engagement sight containing two bodies for battle assessment purposes." Note: CW2 Matthew Warren Pierson was also flying the Kiowa helicopter. Pierson drew a sketch.

204 wearing "all white": *USA v. Lorance,* 679–80.

204 "seven or eight military-age males": CID, Lorance, 7/19/2012; *USA v. Lorance,* 685, 690–91.

205 McNair dropped smoke: *USA v. Lorance,* 685–86.

205 Swanson told the military judge: *USA v. Lorance,* 707–8.

205 "BDA was the next most important thing": *USA v. Lorance,* 707–8.

206 house-to-house search: *USA v. Lorance,* 549.

206 "we would shoot them if they did so": *USA v. Lorance,* 454.

206 "Lieutenant Lorance became irate": *USA v. Lorance,* 455.

206 Lorance asked Shilo: *USA v. Lorance,* 742–43. In this testimony, the defense attorney is asking about testimony Shilo gave in the Article 32 hearing.

207 "I wasn't going to shoot": *USA v. Lorance,* 743–44.

207 "We were intercepting ICOM chatter": *USA v. Lorance,* 511.

207 coming from the grape rows: *USA v. Lorance,* 510–11.

208 Wolfhound chatter: "Memorandum thru Staff Judge Advocate, 82nd Airborne Division, Rule for Courts-Martial 1102, Motion to the

Convening Authority—Newly Discovered Evidence in United States v. Clint A. Lorance," November 30, 2012, author collection.

208 Haggard continued: interview with Dallas Haggard; interview with Michael McGuinness (formerly Herrmann).

209 by a tall tree: *USA v. Lorance,* 452, 548.

209 "We don't necessarily": *USA v. Lorance,* 373.

210 ordered the two soldiers: *USA v. Lorance,* 643–44.

210 Wingo went first: There is a timeline discrepancy here. Shilo testified that the twelve-year-old boy took the motorcycle "immediately after" the initial killing of the two men. This timing is impossible, since the photographs from the attack helicopter show the motorcycle next to the bodies, still with the kickstand down. One must assume this happened later. Wingo and Leon testified: "The motorcycle was stationary on a kickstand, as you can see here. The kickstand was down, keeping the bike up—upright. There were no weapons, no odd wiring to indicate that it was a suspicious vehicle of any kind, sir. We then searched the bodies for any evidence to give to the COIST, Private First Class Skelton."

210 "Myself and Specialist Leon": *USA v. Lorance,* 646.

210 "There were no weapons": *USA v. Lorance,* 651–52.

210 "The local national": *USA v. Lorance,* 545–46, 579–82.

211 "The interpreter was with me": *USA v. Lorance,* 580.

211 "The patrol went through": *USA v. Lorance,* 581.

211 "The locals started to move": *USA v. Lorance,* 652–53.

212 Lorance told Frace to use the radio: *USA v. Lorance,* 553.

212 "Say exactly what I say": *USA v. Lorance,* 630; while exfilling to the strongpoint, another argument ensued. Frace was on the ASIP radio. Lorance became angry. "He got kind of mad at me because I wasn't wording the transmission the way he wanted me to say it. He told me that he knows how to word things the right way, because he used to work in the brigade TOC, that no one would ask any questions."

213 "My next immediate concern": *USA v. Lorance,* 711.

213 "He gave me a shawl": *USA v. Lorance,* 550–51.

213 "I met up with": *USA v. Lorance,* 332.

214 "Soon after the shooting": *USA v. Lorance,* 323–24.

214 "They moved out to the two": *USA v. Lorance,* 324.

214 "First Lieutenant Lorance came in": *USA v. Lorance,* 333–35.

215 "He kept trying": interview with James Twist.

215 "While we were doing that": *USA v. Lorance,* 554. Skelton said: "I would do the storyboard after each mission involving 1 Charlie. I would do the storyboards, and I would do COIST intel analyst comments."

216 "I told him that there were concerns": *USA v. Lorance,* 713–14.

220 "record[ing] data pertaining to": CID, Lorance, 7/5/2012.

220 hands tested positive: *USA v. Lorance,* 554. Skelton also discusses Expray 2 tests on 591–93.

220 "Request for Private Medical History": CID, Lorance, 7/6/2012.

221 "cleansing statement": CID, Lorance, 7/5/2012.

221 "not advised": CID, Lorance, 7/5/2012.

221 "They were like hardass LAPD cops": interview with Michael McGuinness (formerly Herrmann); interview with Anthony Reynoso.

222 "They kept trying": interview with Dallas Haggard.

222 "There are a lot of moving parts": CID, Lorance, 7/6/2012.

223 "Excellent job": CID, Lorance, 1/4/2013.

223 "Upon exiting": CID, Lorance, 7/7/2012.

223 The explanation of this event: correspondence with James Skelton. Whether or not Skelton was acting as a criminal informant for army investigators, he would not say.

223 At ten o'clock that morning: CID, Lorance, Agent's Investigation Report, Crime Scene Examination, July 7, 2012, Exhibits 53–54.

224 "Security concerns imposed": CID, Lorance, Agent's Investigation Report, Crime Scene Examination, July 7, 2012, Exhibit 55.

224 Second Engagement: CID, Lorance, Agent's Investigation Report, Crime Scene Examination, July 7, 2012, Exhibits 53–54.

224 approached by a villager: CID, Lorance, 7/7/2012; Agent's Investigation Report, Crime Scene Examination, July 7, 2012, Exhibit 56.

224 Taskera #1571683: CID, Lorance, 7/7/2012.

224 His name was Abdul Ahad: CID, Lorance, 7/7/2012.

225 Hanna was concerned: CID, Lorance, 7/8/2012; *Commander's Report, Lorance,* 235.

225 "seemed to be unaware": CID, Lorance, 7/8/2012.

225 "After [Lorance] returned": CID, Lorance, 7/15/2012.

225 Pierce told investigators: CID, Lorance, 7/8/2012, 8/9/2012.

225 Lorance's reaction: CID, Lorance, 7/8/2012; *Commander's Report, Lorance,* 236–39.

226 "Haji Karimullah is an important interview": CID, Lorance, 7/15/2012.

226 began working with the Afghan Local Police: CID, Lorance, 7/22/2012. Harrar was part of the STT Task Force Fury at Pasab.

227 The testimony Haji Karimullah gave: CID, Lorance, 7/23/2012, 10:55, Karimullah to Mitchell.

229 payment was commonplace: CID, Lorance, 7/21/2012, 00:25. The CID agents call it a selation payment. I am inferring this is either a typo or an incorrect transliteration of the singular of "solatia," which is "solatium."

229 "Mr. Karimullah stated": CID, Lorance, 7/23/2012.

229 "The one school we built": interview with James Twist.

230 found no probable cause: CID, Lorance, 1/8/2013.

230 Otto gave his legal opinion: CID, Lorance, 1/8/2013.

231 "Local National Victims": CID, Lorance, 1/1/2013.

231 "SAC added the DOB": CID, Lorance, 1/8/2013.

231 "Lieutenant Latino was in the hospital": interview with Samuel Walley.

232 President Obama visited: photograph of President Obama and Walley, Samuel Walley's collection.

233 "Lorance was convicted": John Ramsey, "Army First Lieutenant Found Guilty of Murder, Other Charges for Actions in Afghanistan," *Fayetteville Observer,* August 2, 2013.

237 had committed suicide: interview with John Maher.

239 exhume the bodies: CID, Lorance, 7/13/12.

239 courts component: "Rule of Law Programs in Afghanistan," Department of State, Fact Sheet, May 4, 2012.

239 fifty or sixty times: interview with John Maher.

241 "Bill Carney is *the* expert": Maher, in *Leavenworth*, Episode 3, minute 43.

242 "Bill Carney was the man": Don Brown, *Travesty of Justice: The Shocking Prosecution of Lt. Clint Lorance*, 283.

242 "I sent Bill Carney an email": Don Brown, *Travesty of Justice: The Shocking Prosecution of Lt. Clint Lorance*, 283. John Maher told me this same story in an interview in June 2019.

242 80 percent: CENTCOM, "Conference Maps the Way Ahead for Biometrics in Afghanistan," October 15, 2010.

242 tried and convicted: interview with William Carney. I was not able to independently verify this information; Carney said it was "sensitive."

243 "John [Maher] hoped and prayed: Don Brown, *Travesty of Justice: The Shocking Prosecution of Lt. Clint Lorance*, 283.

243 "John sat": Don Brown, *Travesty of Justice: The Shocking Prosecution of Lt. Clint Lorance*, 283–84.

244 "We spoke": interview with John Maher; see also Don Brown, *Travesty of Justice: The Shocking Prosecution of Lt. Clint Lorance*, 283–85.

244 "burn rights" certified: interview with William Carney; William Carney résumé, "Chief Investigators, William 'Coach' Carney," author collection.

245 "I said, 'Coach'": interview with John Maher.

245 "were a smoking gun": interview with John Maher; Don Brown, *Travesty of Justice: The Shocking Prosecution of Lt. Clint Lorance*, 305.

247 "The most glaring issue": Maher, in *Leavenworth*, Episode 3, minute 43.

248 "An informed electorate": *McGehee v. C.I.A.*, 697 F.2d, 1095, U.S. District Court of Appeals, District of Columbia Circuit, October 4, 1983, 697.

251 "[I] listened to another group": Dr. George E. Tita, "Human Terrain Mapping, Geospatial Intelligence, LAPD—DGI," University of California, Irvine, transcript, February 3, 2012. Tita refers to the U.S.

Army's Human Terrain System Program, a DARPA-conceived program that sent academics into the war theater to map humans.

252 With funding: "Can Math and Science Help Solve Crimes? Scientists Work with Los Angeles Police to Identify and Analyze Crime 'Hotspots,'" *Science Daily,* Science News, University of California, Los Angeles, February 27, 2010; Martin B. Short, P. Jeffrey Brantingham, Andrea L. Bertozzi, and George E. Tita, "Dissipation and Displacement of Hotspots in Reaction-Diffusion Models of Crime," *PNAS,* February 22, 2010.

253 "A lot of human behavior": Ellen Huet, "Server and Protect: Predictive Policing Firm PredPol Promises to Map Crime Before It Happens," *Forbes,* February 11, 2015.

253 "Criminals want to replicate": "Dr. George Mohler: Mathematician and Crime Fighter," DataSmart, City Solutions, Harvard Kennedy School Ash Center for Democratic Governance and Innovation, May 8, 2013. He said, "If you can deter a crime from happening, that's one less person who has to go through the criminal justice system."

253 "hunter-gatherers": "Can Math and Science Help Solve Crimes? Scientists Work with Los Angeles Police to Identify and Analyze Crime 'Hotspots,'" *Science Daily,* Science News, University of California, Los Angeles, February 27, 2010.

253 "Can we actually": "Can Math and Science Help Solve Crimes? Scientists Work with Los Angeles Police to Identify and Analyze Crime 'Hotspots,'" *Science Daily,* Science News, University of California, Los Angeles, February 27, 2010.

253 "Predictive policing is": Andrew G. Ferguson, "Policing Predictive Policing," *Washington University Law Review* 94, no. 5 (2017).

254 "[These] pretrial risk assessment tools": "Mapping Pretrial InJustice Database + Webinar: Predictive Policing," YouTube, February 13, 2020, https://www.youtube.com/watch?v=FvInn7NnZPs, accessed March 1, 2020.

254–55 "broken windows" policing: Nathan Munn, "This Predictive Policing Company Compares Its Software to 'Broken Windows' Policing," Motherboard: Tech by *Vice,* June 11, 2018; Caroline Haskins,

"Academics Confirm Major Predictive Policing Algorithm Is Fundamentally Flawed," *Vice,* February 14, 2019.

255 "humans must always": interview with Patrick Biltgen.

255 1.8-gigapixel surveillance drone: "Rise of the Drones," PBS *Nova,* January 23, 2013.

255 "It is important": "Rise of the Drones," PBS *Nova,* January 23, 2013.

255 "the devil is in the data": Vincent Southerland, "With AI and Criminal Justice, the Devil Is in the Data," ACLU.org, April 9, 2018.

255 "PredPol is currently": a PredPol overview is available at the company's website, PredPol.com.

257 "Caution: Data is only as reliable": New Systems Training and Integration Office, *Introduction to Biometrics and Biometric Systems,* July 7, 2011.

258 Carney's sworn affidavit: *Affidavit of William C. Carney,* State of Illinois, County of Cook, August 31, 2015. In the affidavit, Carney swore: "the testimony I present as part of the Petition for a New Trial in the matter of *U.S. v. Clint Lorance,* is the truth, the whole truth, and nothing but the truth, so help me God."

260 lost the war: Sarah Almukhtar and Rod Nordland, "What Did the U.S. Get for $2 Trillion in Afghanistan?" *New York Times,* December 9, 2019.

260 "Our goal is to help reduce": interview with Ross McNutt. All quotes are from our interviews unless noted otherwise.

260 Angel Fire: interview with Ross McNutt.

261 a compelling event: interview with Ross McNutt.

261 PSS cameras captured: www.pss-1.com, see "Public Safety." The 2012 digital map of the driver's crime spree resembles a 2012 GEOINT product from Afghanistan.

261 trapezoidal footprint: Patrick Biltgen and Stephen Ryan, *Activity-Based Intelligence: Principles and Applications,* 169.

262 "You know where": Craig Timberg, "New Surveillance Technology Can Track Everyone in an Area for Several Hours at a Time," *Washington Post,* February 5, 2014.

262 judge's delivery of the not-guilty verdict: Monte Reel, "Secret Cameras Watch Baltimore's Every Move from Above," *Bloomberg Businessweek,* August 23, 2016.

262 filed a lawsuit against the Baltimore Police: "ACLU Challenges Pilot Aerial Surveillance Baltimore," ACLU press release, April 9, 2020.

262 "A single camera": Craig Timberg, "New Surveillance Technology Can Track Everyone in an Area for Several Hours at a Time," *Washington Post,* February 5, 2014.

264 soda from McDonald's: Edgar Sandoval, "N.Y.P.D. to Remove DNA Profiles of Non-Criminals from Database," *New York Times,* February 20, 2020.

265 readers expressed shock: Jan Ransom and Ashley Southall, "N.Y.P.D. Detectives Gave a Boy, 12, a Soda. He Landed in a DNA Database," *New York Times,* August 15, 2019.

266 solved using DNA: Joseph Wambaugh, *The Blooding: The True Story of the Narborough Village Murders,* 153.

266 twenty-four months: interview with Dr. Richard Selden. See also John E. Siedlarz et al., *Biometric Identification Verification Technology Status and Feasibility Study,* DNA-TR-94-6, Defense Nuclear Agency, September 1994.

267 1.2 million service members: Douglas J. Gillert, "Who Are You? DNA Registry Knows," American Forces Press Service, July 13, 1998.

267 DNA reference card: Armed Forces Medical Examiner, "8 Millionth DNA Reference Card Processed," Armed Forces Medical Examiner, Dover Air Force Base, press release, March 29, 2019.

267 Black Helix: DoD, Office of the Under Secretary of Defense for Acquisition, Technology, and Logistics, *Report of the Defense Science Board Task Force on Defense Biometrics,* March 2007, 31.

267 30,000 new samples: DoD, Office of the Under Secretary of Defense for Acquisition, Technology, and Logistics, *Report of the Defense Science Board Task Force on Defense Biometrics,* March 2007, 32.

268 two samples every three days: Steven Fidler, "Fears over Covert DNA Database," *Financial Times,* November 17, 2008.

268 inspired young Richard Selden: interview with Dr. Richard Selden, here and throughout this chapter.

271 website called GEDmatch: Margaret Press, PhD, and Colleen Fitzpatrick, PhD, "The DNA DOE Project, Our First Two Years of Success," presentation, International Symposium on Human Identification, September 26, 2019.

272 Uighur descent: "China: Minority Region Collects DNA from Millions, Private Information Gathered by Police, Under Guise of Public Health Program," Human Rights Watch, December 13, 2017.

272 reeducation program: Lindsay Maizland, "China's Repression of Uighurs in Xinjiang," Council on Foreign Relations, November 25, 2019.

273 "I had thought we were sharing": Sui-Lee Wee, "China Uses DNA to Track Its People, with the Help of American Expertise," *New York Times,* February 21, 2019; see also Marisa Peryer and Serena Cho, "Yale Geneticist Provided Data Used for Chinese Surveillance of Uighurs," *Yale News,* February 22, 2019.

273 probabilistic genotyping: Samantha O. Wandzek, "Courtroom Testimony for Probabilistic Genotyping," International Symposium on Human Identification, November 29, 2017.

273 Sacramento County coroner: interview with Kimberly D. Gin, Sacramento County coroner; Department of Homeland Security, "Snapshot: S&T's Rapid DNA Technology Identified Victims of California Wildfire," newsroom, DHS.gov, April 23, 2019; interview with Dr. Richard Selden.

274 otherwise "unrecognizable": Department of Homeland Security "Snapshot: Rapid DNA Identifies Conception Boat Fire Victims," newsroom, DHS.gov, December 10, 2019. DHS S&T claimed it financed all of Selden's work. As per Selden, this is not accurate.

274 took on darker overtones: Nick Miroff, "Homeland Security to Test DNA of Families at Border in Cases of Suspected Fraud," *Washington Post,* May 1, 2019; Caitlin Dickerson, "U.S. Government Plans to Collect DNA from Detained Immigrants," *New York Times,* October

2, 2019; Taylor Dolven, "The U.S. Is Still Holding 'About 100' Kids Under 5 Who Were Taken from Parents at Border," Vice News, July 5, 2018; interview with Dr. Richard Selden.

274 Republican members of Congress: Garret Graves et al., "Dear Mr. President . . ." Congress of the United States, Washington, DC, 20515, March 1, 2019.

275 "Staff Sergeant Israel P. Nuanes": Don Brown, *Travesty of Justice: The Shocking Prosecution of Lt. Clint Lorance*, 309.

276 "War is hard": Dave Philipps, "Cause Célèbre, Scorned by Troops," *New York Times*, February 24, 2015.

278 "He was recovering nicely": interview with Todd Fitzgerald.

278 he remained optimistic: interview with Daniel Williams.

278 twenty-one-gun salute: Ivey DeJesus, "Military Veteran Cpl. Matthew Hanes Honored and Remembered," Penn Live, August 14, 2015.

279 By accident he fired: interview with Samuel Walley.

280 Ruhl was shot in the abdomen: "Coroner: Victim Shot During Struggle with Gun Dies at Hospital," Wane.com, March 28, 2016.

281 Writ of Habeas Corpus: interview with John Maher, here and throughout this chapter.

281 "Most of our donations": interview with Maj. Bill Donahue.

281 "That night": Maher, in *Leavenworth*, Episode 4, minute 43.

282 appealed to President Trump: John N. Maher, "Dear Mr. President, Mr. Pardon Attorney, and Mr. Secretary: On Behalf of First Lieutenant Clint A. Lorance," June 1, 2018.

283 What biometric data: Of note: Secretary of the Army Mark Esper was included in the "Request for Presidential Action—Disapprove Convictions." Prior to Esper's position as the head of the U.S. Army, he served as vice president for government relations at the Raytheon Company, the defense contractor in charge of "biometric-related services and support" for the Biometrics Task Force from 2009 to 2014. In July 2019, Esper was sworn in as the twenty-seventh secretary of defense; clearly he was well versed in biometrics when he advised

President Trump on the Lorance pardon. See "Raytheon to Supply Biometrics to DOD," Raytheon press release, January 28, 2009.

283 IED Event #12/1229: Don Brown, *Travesty of Justice: The Shocking Prosecution of Lt. Clint Lorance,* 309; "Member of the Lorance Legal Team: We Need President Trump to Take Action, Donald Brown Says Army 1st Lt. Clint Lorance Is Not Guilty of War Crimes," *Fox & Friends,* April 13, 2019.

285 another group of FOIA documents: *Karimullah Abdul Karim, BID: B2JK4-G7D7.*

286 records for "Ghamai": *Ghamai Mohammad Nabi, BID: B2JK9-B3R3.*

288 "deep state": Nicholas Wu and John Fritze, "Trump Pardons Service Members in High Profile War Crimes Cases," *USA Today,* November 15, 2019.

288 "To be honest with you": "Clint Lorance LIVE in Studio," *The Sean Hannity Show,* November 18, 2019, minute 16:50.

288 from Helmand: interview with William Carney. It's impossible for me to know if Carney really was in Helmand, Afghanistan, when we spoke, but he said he was.

289 filmmakers tracked down: I interviewed one of the *Leavenworth* producers (who asked not to be named), who described to me how their team used an Afghan reporter to conduct these interviews.

291 "One of his victims": interview with Lt. Col. Jay Morse (ret.), lead army prosecutor; see also AR 15-6 SSG Bales Incident.

291 "Fingerprint/DNA": SCOTUS, *Robert Bales v. United States,* On Petition for a Writ of Certiorari to the United States Court of Appeals for the Armed Forces, No. 17, Washington, DC, May 16, 2018, 19.

292 "a burgeoning 'National Surveillance State'": Margaret Hu, "Biometric Cyberintelligence and the Posse Comitatus Act," *Emory Law Journal* 66 (2017): 702–7; see also Margaret Hu, "Big Data Blacklisting," *Florida Law Review* 67, no. 5, Article 5 (March 2016): 1–76.

292 largest per capita share: Liza Lin and Newley Purnell, "A World with a Billion Cameras Watching You Is Just Around the Corner," *Wall Street Journal,* December 6, 2019. The report is from industry researcher IHS Markit.

292 less than one second: interview with Bill Whalen. As police chief, Whalen has concerns about facial recognition in policing, he told me, and the Clearview demonstration test was his idea.

293 no-bid contract: Erin Banco and Spencer Ackerman, "Team Trump Turns to Peter Thiel's Palantir to Track Virus," Daily Beast, April 21, 2020.

293 "Given how tight-lipped": interview with Lauren Zabierek.

293 company website reads: Andy Greenberg, "How a 'Deviant' Philosopher Built Palantir, a CIA-Funded Data-Mining Juggernaut," *Forbes,* August 14, 2013.

294 HHS's Office of Refugee Resettlement: Kevin K. McAleenan et al., "Memorandum of Agreement Among the Office of Refugee Resettlement of the U.S. Department of Health and Human Services, and U.S. Immigration and Customs Enforcement and U.S. Customs and Border Protection of the U.S. Department of Homeland Security Regarding Consultation and Information Sharing Matters in Unaccompanied Alien Children Matters," April 13, 2018.

294 wrote to HHS Secretary Alex Azar: Reed Albergotti, "Lawmakers Call for More Transparency in Health Agency's Pandemic Data Collection Practices," *Washington Post,* July 1, 2020; "Congressional Hispanic Caucus Members Demand Trump Administration Release HHS Contracts with Palantir, Tech Companies," Congressional Hispanic Caucus press release, June 25, 2020.

295 "Lockheed built ABIS": interview with Tom Bush. In October 2020, the DoD announced plans to move ABIS to a cloud environment hosted by Amazon.

296 "persistent near-real-time": Special Operations Forces Acquisition, Technology, and Logistics, Directorate of Science and Technology (SOF AT&L-ST), Broad Agency Announcement, USSOCOM-BAAST-2020, unclassified, 4.0-4.1.1.1.1.2.

296 "rapid, portable, handheld DNA": Special Operations Forces Acquisition, Technology, and Logistics, Directorate of Science and Technology (SOF AT&L-ST), Broad Agency Announcement, USSOCOM-BAAST-2020, unclassified, 4.0-4.1.1.1.1.2.

296 What will society become?: The following individuals and organizations declined to be interviewed for this book: the White House; Secretary of Defense Mark Esper; Secretary of the Army Ryan D. McCarthy; U.S. Army Chief of Public Affairs, Brig. Gen. Amy E. Hannah; U.S. Department of Justice, Office of the Pardon Attorney; P. Jeffrey Brantingham; George E. Tita; and George Mohler.

BIBLIOGRAPHY

BOOKS

Addleton, Jonathan S. *The Dust of Kandahar: A Diplomat Among Warriors in Afghanistan.* Annapolis, MD: Naval Institute Press, 2016.

Biltgen, Patrick, and Stephen Ryan. *Activity-Based Intelligence: Principles and Applications.* Norwood, MA: Artech House, 2016.

Brown, Don. *Travesty of Justice: The Shocking Prosecution of Lt. Clint Lorance.* Denver, CO: Wildblue Press, 2019.

Chandrasekaran, Rajiv. *Little America: The War Within the War for Afghanistan.* New York: Knopf, 2012.

Coll, Steve. *Directorate S: The C.I.A. and America's Secret Wars in Afghanistan and Pakistan.* New York: Penguin, 2018.

Demarest, Geoff. *Winning Insurgent War: Back to Basics.* 2nd Ed. Fort Leavenworth, KS: Foreign Military Studies Office, 2011.

Faulds, Henry. *A Manual of Practical Dactylography: A Work for the Use of Students of the Finger-Print Method of Identification.* London: The Police Review, 1923.

Foucault, Michel. *Abnormal: Lectures at the Collège de France 1974–1975.* Translated by G. Burchell. New York: Picador, 2004.

———. *Discipline and Punish: The Birth of the Prison.* Translated by A. Sheridan. New York: Random House, 1977.

Galton, Francis, F.R.S. *Finger Prints.* London: MacMillan & Co., 1892.

Henry, E. R. *Classification and Uses of Fingerprints.* London: Routledge & Sons, 1900.

Herschel, W. J. *The Origin of Finger-Printing.* London: Oxford University Press, 1916.

Jacobsen, Annie. *Area 51: An Uncensored History of America's Top Secret Military Base.* New York: Little, Brown & Company, 2011.

———. *The Pentagon's Brain: An Uncensored History of DARPA, America's Top Secret Military Research Agency.* New York: Little, Brown & Company, 2015.

Jain, Anil K., Arun A. Ross, and Karthik Nandakumar. *Introduction to Biometrics.* New York: Springer, 2011.

McDermid, Val. *Forensics: What Bugs, Burns, Prints, DNA, and More Tell Us About Crime.* New York: Grove Press, 2014.

Michel, Arthur Holland. *Eyes in the Sky: The Secret Rise of Gorgon Stare and How It Will Watch Us All.* Boston, MA: Houghton Mifflin Harcourt, 2019.

Rashid, Ahmed. *Descent into Chaos: The U.S. and the Disaster in Pakistan, Afghanistan, and Central Asia.* New York: Penguin, 2008.

———. *Taliban: Militant Islam, Oil and Fundamentalism in Central Asia.* New Haven: Yale University Press, 2001.

Shannon, Paul. *Songs of Iraq: A Year Long Deployment.* (Self-published.) 2019.

Steward, Julian H. *Alfred Kroeber, 1876–1960: A Biographical Memoir.* Washington, DC: National Academy of Sciences, 1962.

Toy, Ed. *Pressure Plate: A Perspective on Counter IED Operations in Southern Afghanistan 2008–2009.* XLibris, 2013.

U.S. Army. *The Army Lawyer: A History of the Judge Advocate General's Corps, 1775–1975.* Washington, DC: U.S. Government Printing Office, 1975.

Wambaugh, Joseph. *The Blooding: The True Story of the Narborough Village Murders.* New York: William Morrow, 1989.

Woodward, John D., Jr., Nicholas M. Orlans, and Peter T. Higgins. *Biometrics.* Berkeley, CA: McGraw-Hill/Osborne, 2003.

Zaeef, Abdul Salam. *Ambassador P.O.W.* Johannesburg, South Africa: Cage, 2015.

———. *My Life with the Taliban.* London: Hurst, 2011.

MONOGRAPHS AND REPORTS

Air Land Sea Application Center. *Biometrics: Multi-Service Tactics, Techniques, and Procedures for Tactical Employment of Biometrics in Support of Operations.* Multi-Service Tactics, Techniques, and Procedures (MTTP), April 2014.

———. *Biometrics: Multi-Service Tactics, Techniques, and Procedures for Tactical Employment of Biometrics in Support of Operations.* Multi-Service Tactics, Techniques, and Procedures (MTTP), May 6, 2016.

Army Criminal Investigation Command (CID) Annual Historical Review 2013–2014, 2015. U.S. Army Criminal Investigation Command Crime Records Center, Quantico, VA, September 15, 2016.

Benard, Cheryl et al. *The Battle Behind the Wire: U.S. Prisoner and Detainee Operations from World War II to Iraq.* Office of the Secretary of Defense, National Defense Research Institute, 2011.

Brown, Katherine Ann. *Patterns in the Chaos: News and Nationalism in Afghanistan, America and Pakistan During Wartime, 2010–2012.* PhD Dissertation. Columbia University, New York, 2013.

Cavanaugh, Michael. *The Evolution of the Use of Biometrics by the Department of Defense.* November 20, 2017.

Coburn, Noah, and John Dempsey. "Informal Dispute Resolution in Afghanistan." Special Report 247. Washington, DC: United States Institute of Peace, August 2010.

Dirks, Emile, and James Leibold. *Genomic Surveillance: Inside China's DNA Dragnet.* Policy Brief Report No. 34/2020. International Cyber Policy Centre, Australian Strategic Policy Institute, June 2020.

Dorronsoro, Gilles. *The Taliban's Winning Strategy in Afghanistan.* Washington, DC: Carnegie Endowment for International Peace, 2009.

Electronic Frontier Foundation. "Mandatory National IDs and Biometric Databases." EFF.org (n.d.).

Elsea, Jennifer. *The Posse Comitatus Act and Related Matters: The Use of the Military to Execute Civilian Law.* R42659, Ver. 7. Congressional Research Service, November 6, 2018.

Federal Bureau of Investigation, Bomb Data Center. *General Information Bulletin 97-1.* 1997.

"Final Report of the Commission on the Intelligence Capabilities of the United States Regarding Weapons of Mass Destruction." Report to the President of the United States, March 31, 2005.

Forsberg, Carl. "Counterinsurgency in Kandahar: Evaluating the 2010 Hamkari Campaign." *Afghanistan Report 7.* Washington, DC: Institute for the Study of War, December 2010.

Green, Lt. Col. Michael R. *Zeroing Biometrics: Collecting Biometrics Before the Shooting Starts.* MA Thesis. Joint Forces Staff College, Joint Advanced Warfighting School, Norfolk, VA, April 26, 2012.

International Security Assistance Force (ISAF) Joint Command (IJC). *IED Identification: Ghamai Mohammad Nabi, BID: B2JK9-B3R3.* Biometric Intelligence Program. United States Army Intelligence and Security Command (includes multiple declassification releases).

———. *IED Identification: Karimullah Abdul Karim, BID: B2JK4-G7D7.* Biometric Intelligence Program. United States Army Intelligence and Security Command (includes multiple declassification releases).

Jadoon, Amira. *Allied & Lethal: Islamic State Khorasan's Network and Organizational Capacity in Afghanistan and Pakistan.* Combating Terrorism Center at West Point. West Point, NY: United States Military Academy, December 2018.

Joint Chiefs of Staff. "Joint Publication 2-0: *Joint Intelligence.*" October 22, 2013.

———. "Joint Publication 3-05: *Special Operations.*" July 16, 2014.

———. "Joint Publication 3-42: *Joint Explosive Ordnance Disposal.*" September 9, 2016.

Jones, Dalton, and Duane Blackburn. *Identity Intelligence: From Reactionary Support to Sustained Enabler.* Defense Intelligence Agency, Identity Intelligence Project Office (I2PO), August 21, 2012.

Llinas, James, and James Scrofani. *Foundational Technologies for Activity-Based Intelligence—A Review of the Literature.* NPS-EC-14-001. Monterey, CA: Naval Postgraduate School, February 2014.

MacDonald, Major Erick. "Weak Signals: Identity Intelligence and Its Impact on the Future Operation Environment." North York, ON: Canadian Forces College, May 2019.

Mason, R. Chuck. "Status of Forces Agreement (SOFA): What Is It, and How Has It Been Utilized?" Congressional Research Service, March 15, 2012.

Meyerle, Jerry, and Carter Malkasian. *Insurgent Tactics in Southern Afghanistan: 2005–2008.* Alexandria, VA: Marine Corps Intelligence Activity, Center for Naval Analyses (CNA), August 2009.

Moyar, Mark. *Village Stability Operations and the Afghan Local Police.* Report 14-7. MacDill AFB, FL: Joint Special Operations University, October 2014.

Muller, Benjamin, and John Measor. "Securitizing the Global Norm of Identity Biometrics and *Homo Sacer* in Fallujah." Queen's University Belfast, N. Ireland, May 20, 2005.

National Institute of Justice. *The Fingerprint Sourcebook.* Washington, DC: U.S. Department of Justice, Office of Justice Programs (n.d.).

National Science and Technology Council. *Biometrics in Government Post-9/11: Advancing Science, Enhancing Operations.* Washington, DC: Office of Science and Technology, September 11, 2008.

———. Committee on Technology. Committee on Homeland and National Security, Subcommittee on Biometrics. *Biometrics History.* March 31, 2006.

National System for Geospatial Intelligence. *Geospatial Intelligence (GEOINT) Basic Doctrine.* Publication 1.0. Approved for Public Release 18-142. April 2018.

North Atlantic Treaty Organization, International Security Assistance Force. *NATO Rule of Law Field Support Mission (NROLFSM).* June 9, 2011.

Office of the NGA Historian. "The Advent of the National Geospatial-Intelligence Agency." September 2011.

Ohman, Maj Mynda G. *Integrating Title 18 War Crimes into Title 10: A Proposal to Amend the Uniform Code of Military Justice.* ML Thesis. George Washington University Law School, Washington, DC, May 22, 2005.

Pedersen, Janni. *The Symbolic Mind: Apes, Symbols and the Evolution of Language.* PhD Dissertation. Iowa State University, Ames, IA, 2012.

Perry, Walter L., Brian McInnis, Carter C. Price, Susan C. Smith, and John S. Hollywood. *Predictive Policing: The Role of Crime Forecasting in Law Enforcement Operations.* RR-233-NIJ. Santa Monica, CA: RAND Corporation, 2010.

Peters, Gretchen. "Crime and Insurgency in the Tribal Areas of Afghanistan and Pakistan." Harmony Project, Combating Terrorism Center at West Point. West Point, NY, October 15, 2010.

Price, Maj. Jerry S. *History of the United States Disciplinary Barracks, 1875–Present.* Student Study Project #4016. Fort Leavenworth, KS: United States Army Command and General Staff College, May 7, 1978.

Pruett, Richard K. *Identification—Friend or Foe? The Strategic Uses and Future Implications of the Revolutionary New ID Technologies.* MA Thesis. U.S. Army War College, Carlisle, PA, March 15, 2006.

Rizvi, Syed A., P. Jonathon Phillips, and Hyeonjoon Moon. *The FERET Verification Testing Protocol for Face Recognition Algorithms.* NISTIR 6281. Gaithersburg, MD: U.S. Department of Commerce, Technology Administration, National Institute of Standards and Technology, October 1998.

Shontz, Douglas. *DNA as Part of Identity Management for the Department of Defense.* Santa Monica, CA: RAND Corporation, 2010.

Shrout, Lt. Col. Michael S. *Biometrically Supported Census Operations as a Population Control Measure in Counterinsurgency: A Monograph.* MA Thesis. School of Advanced Military Studies, Command and General Staff College, Fort Leavenworth, KS, 2011.

Siedlarz, John E. et al. *Biometric Identification Verification Technology Status and Feasibility Study.* DNA-TR-94-6. Alexandria, VA: Defense Nuclear Agency, September 1994.

Sonderegger, Bernhard, and Martin Urs Peter. *The Fingerprint: 100 Years in the Service of the Swiss Confederation.* Bern, Switzerland: Federal Office of Police (FEDPOL), November 2012.

Special Inspector General for Afghanistan Reconstruction. *2019 High-Risk List.* SIGAR 19-25-HRL. Arlington, VA, April 2020.

———. *Department of Defense: More than 75 Percent of All SIGAR Audit and Inspection Report Recommendations Have Been Implemented.* SIGAR 15-29-AR. Arlington, VA, January 2015.

———. *Department of Defense's Efforts to Maintain, Operate, and Sustain the Afghan Automated Biometrics Identification System: Audit of Costs Incurred by Ideal Innovations Inc.* SIGAR 20-32 Financial Audit. Arlington, VA, April 2020.

———. *Improvised Explosive Devices: Unclear Whether Culvert Denial Systems to Protect Troops Are Functioning or Were Ever Installed.* SIGAR SP-13-8. Arlington, VA, July 2013.

———. *Justice Center in Parwan Courthouse: Poor Oversight Contributed to Failed Project.* SIGAR 14-7 Inspection Report. Arlington, VA, October 2013.

———. *Rule of Law in Afghanistan: U.S. Agencies Lack a Strategy and Cannot Fully Determine the Effectiveness of Programs Costing More Than $1 Billion.* SIGAR 15-68 Audit Report. Arlington, VA, July 2015.

Stop LAPD Spying Coalition. *Before the Bullet Hits the Body: Dismantling Predictive Policing in Los Angeles.* Los Angeles, May 8, 2018.

"Summary Report of DoD Funded Lighter-Than-Air-Vehicles." Dates Covered: 2012. Washington, DC: Office of the Assistant Secretary of Defense for Research and Engineering, Rapid Reaction Technology Office, October 2015.

Teeple, Nancy. *Canada in Afghanistan: 2001–2010: A Military Chronology.* Defence R&D Canada, CORA CR 2010-282. Ottawa, ON, December 2010.

Thomas, Clayton. *Afghanistan: Background and U.S. Policy in Brief.* R45122. Congressional Research Service. August 1, 2019.

Tribute to Combined Joint Interagency Task Force 435. Congressional Record Vol. 160, No. 134. September 18, 2014.

United Nations Counter-Terrorism Centre and the Biometrics Institute. *United Nations Compendium of Recommended Practices: For the Responsible Use & Sharing of Biometrics in Counter Terrorism* (n.d.).

United States Geospatial Intelligence Foundation. *The State and Future of GEOINT Report.* 2018.

———. *The State and Future of GEOINT Report.* 2019.

United States Military Academy. Simon Center for the Professional Military Ethic. *Character Program (Goldbook), Academic Year 2020.* West Point, NY, 2019.

U.S. Army. "Commander's Guide to Money as a Weapons System: Tactics, Techniques, and Procedures." Handbook No. 09-27. U.S. Army Combined Arms Center, April 2009.

U.S. Army Combined Arms Center. *Commander's Guide to Biometrics in Afghanistan: Observations, Insights, and Lessons.* Handbook No. 11–25. Fort Leavenworth, KS: Center for Army Lessons Learned, April 2011.

———. *Company Intelligence Support Team (COIST) Handbook: Tactics, Techniques, and Procedures.* No. 10-20. Fort Leavenworth, KS: Center for Army Lessons Learned, January 2010.

U.S. Army Intelligence Center of Excellence. "Biometrics Automated Toolset-Army (BAT-A) STRAP Increment 1" (UPDATE) (version 3.0) (n.d.).

US Army and Marine Corps Counterinsurgency Field Manual. 2007.

U.S. Army National Ground Intelligence Center. *BMAT Biometric Intelligence Analysis Report: Abdulhat Abdulhat, BID: B28JM-UUYZ.* United States Army Intelligence and Security Command, March 18, 2013 (includes multiple declassification releases).

———. *BMAT Biometric Intelligence Analysis Report: Mohammad Rahim Sadozay, BID: B28JP-QWTY.* United States Army Intelligence and Security Command, June 22, 2012 (includes multiple declassification releases).

U.S. Department of Defense. *Report on Progress Toward Security and Stability in Afghanistan.* Washington, DC, December 2012.

U.S. Department of Defense, Biometrics Task Force. *Biometrics Task Force Annual Report FY07.* Washington, DC, 2007.

———. *Biometrics Task Force Annual Report FY08.* Washington, DC, 2008.

———. *Biometrics Task Force Annual Report FY09.* Washington, DC, 2009.

U.S. Department of Defense, Defense Forensics and Biometrics Agency. *Broad Agency Announcement for Basic, Applied & Advanced Scientific Research.* U.S. Army Contracting Command–Aberdeen Proving Ground, Research Triangle Park, NC, June 30, 2015.

———. *Forensic Science Lexicon.* January 2018.

U.S. Department of Defense, Defense Science Board. *Defense Science and Technology.* Washington, DC: Office of the Under Secretary of Defense for Acquisition, Technology, and Logistics, May 2002.

U.S. Department of Defense, Director, Operational Test & Evaluation. *Automated Biometric Identification System (ABIS) Version 1.2: Initial Operational Test and Evaluation Report.* May 2015.

U.S. Department of Defense, Inspector General. *Semiannual Report to Congress: October 1, 2012, to March 31, 2013.* Alexandria, VA, 2013.

U.S. Department of Defense, Joint Chiefs of Staff. *Counterinsurgency Operations.* Joint Publication 3-24. October 5, 2009.

———. *Geospatial Intelligence in Joint Operations.* Joint Publication 2-03. July 5, 2017.

———. *Identity Activities.* Joint Doctrine Note 2-16. August 3, 2016.

———. *Multinational Operations.* Joint Publication 2-16. July 16, 2013.

U.S. Department of Defense, Multi-National Force West. *Specific Modified Training Techniques and Procedures (TTP's).* Ver. 1, Pt. 2. June 2007.

U.S. Department of Defense, National Reconnaissance Office. *Sentient Program.* REL to USA, FVEY. DECL ON 25X1, 20670112. INCG 1.0, February 13, 2012.

U.S. Department of Defense, Office of the Inspector General. *Review of DoD-Directed Investigations of Detainee Abuse (U).* Report No. 06-INTEL-10. Arlington, VA, August 25, 2006.

U.S. Department of Defense, Office of the Under Secretary of Defense for Acquisition, Technology, and Logistics. *Report of the Defense Science Board Task Force on Defense Biometrics.* Washington, DC, March 2007.

U.S. Department of Defense, Research and Engineering. Defense Biometrics and Forensics. *Defense Biometric and Forensic Office Research Development, Test and Evaluation Strategy.* Washington, DC, January 6, 2015.

U.S. Department of State and the Broadcasting Board of Governors Office of Inspector General. *Report of Inspection: Rule-of-Law Programs in Afghanistan.* Report No. ISP-I-08-09. January 2008.

U.S. Department of State, Office of the Coordinator for Counterterrorism. *Country Reports of Terrorism 2010.* August 2011.

U.S. Government Accountability Office. *Bioforensics: DHS Needs to Conduct a Formal Capability Gap Analysis to Better Identify and Address Gaps.* Report to Congressional Requesters. GAO-17-177. Washington, DC, January 2017.

———. *Defense Acquisitions: Future Aerostat and Airship Investment Decisions Drive Oversight and Coordination Needs.* Report to the Subcommittee on Emerging Threats and Capabilities, Committee on Armed Services, U.S. Senate. GAO-13-81. Washington, DC, October 2011.

———. *Defense Biometrics: Additional Training for Leaders and More Timely Transmission of Data Could Enhance the Use of Biometrics in Afghanistan.* Report to Congressional Requesters. GAO-12-442. Washington, DC, April 2012.

———. *Defense Biometrics: DOD Can Better Conform to Standards and Share Biometric Information with Federal Agencies.* Report to Congressional Requesters. GAO-11-276. Washington, DC, March 2011.

———. *DoD Biometrics and Forensics: Progress Made in Establishing Long-term Deployable Capabilities, but Further Actions Are Needed.* Report to Congressional Committees. GAO-17-580. Washington, DC, August 2017.

———. *Warfighter Support: Actions Needed to Improve Explosive Ordnance Disposal Forces Planning.* Report to Congressional Committees. GAO-19-698. Washington, DC, September 2019.

U.S. Joint Forces Command. *Commander's Handbook for Attack the Network.* Ver. 1.0. Suffolk, VA: Joint Warfighting Center, Joint Doctrine Support Division, May 20, 2011.

U.S. Joint Targeting School Staff, J-7. *Joint Battle Damage Assessment Course Syllabus.* Virginia Beach, VA: Joint Targeting School, October 2015.

Voelz, Glenn J. *The Rise of iWar: Identity, Information, and the Individualization of Modern Warfare.* U.S. Army War College, Strategic Studies Institute, Carlisle, PA, October 2015.

The White House. *Report on the Legal and Policy Frameworks Guiding the United States' Use of Military Force and Related National Security Operations.* Washington, DC, December 2016.

Woodward, John D., Jr., Christopher Horn, Julius Gatune, and Aryn Thomas. *Biometrics: A Look at Facial Recognition*. Santa Monica, CA: RAND Public Safety and Justice for the Virginia State Crime Commission, 2003.

Woodward, John D., Jr. et al. *Army Biometric Applications: Identifying and Addressing Sociocultural Concerns*. Santa Monica, CA: RAND Corporation, 2001.

Wyler, Liana Sun, and Kenneth Katzman. *Afghanistan: U.S. Rule of Law and Justice Sector Assistance*. R41484. Congressional Research Service, Report for Congress. November 9, 2010.

ARTICLES

Abrahams, Edward, Patrick T. Biltgen, Peter Hanson, and Shannon C. Pankow. "Everything, Everywhere, All the Time—Now What?" *Trajectory Magazine*, February 1, 2018.

"ACLU Challenges Pilot Aerial Surveillance Baltimore." ACLU press release, April 9, 2020.

Adair, Staff Sgt. Christopher. "Intelligence Support Teams' Support to Logistics Organizations." Army.mil, January 13, 2014.

Aftergood, Steven. "Identity Intelligence and Special Operations." FAS.org, July 30, 2014.

"Agency U-2 Pilots: Hervey Stockman." From the Vault, CIA.gov.

Albergotti, Reed. "Lawmakers Call for More Transparency in Health Agency's Pandemic Data Collection Practices." *Washington Post*, July 1, 2020.

Alderton, Matt. "Safe + Found: Geospatial Applications Hold Great Utility for Public Safety Planning and Response." TRJ-022. *Trajectory Magazine* 2 (2014).

Almukhtar, Sarah, and Rod Nordland. "What Did the U.S. Get for $2 Trillion in Afghanistan?" *New York Times*, December 9, 2019.

Armed Forces Medical Examiner. "8 Millionth DNA Reference Card Processed." Armed Forces Medical Examiner, Dover Air Force Base, press release, March 29, 2019.

Ashton, Adam. "Report: Quran Burnings Revealed Distrust at Afghanistan's Bagram Jail." *Tacoma News Tribune,* August 31, 2012.

Atwood, Chandler P. "Activity-Based Intelligence: Revolutionizing Military Intelligence Analysis." *Joint Force Quarterly* 77 (2nd Quarter 2015): 24–33.

Ballantyne, Michael, Robert S. Boyer, and Larry Hines. "Woody Bledsoe: His Life and Legacy." *AI Magazine* 17, no. 1 (Spring 1996).

Banco, Erin, and Spencer Ackerman. "Team Trump Turns to Peter Thiel's Palantir to Track Virus." Daily Beast, April 21, 2020.

Bell, Suzanne et al. "A Call for More Science in Forensic Science." *PNAS* 115, no. 18 (May 2018): 4541–44.

Bergen, Peter. "The Crossroads: Can We Win in Afghanistan?" *The New Republic,* May 3, 2011.

Bhattacharya, Shaoni. "Fast-Track DNA Tests Confirm Saddam's Identity." *New Scientist,* December 15, 2003.

Billing, Lynzy. "Inside Afghanistan's Main Forensic Lab." *Los Angeles Times,* August 28, 2019.

Bowcott, Owen. "UK: FBI Wants Instant Access to British Identity Data." *The Guardian,* January 15, 2008.

Brewster, Murray. "'Au Revoir, Zangabad,' Canadian Army Hands Over Afghan Village to U.S." Canadian Press, June 2011.

Bureau of Alcohol, Tobacco, Firearms and Explosives (ATF). "ATF, Military, Expand Their Collaboration Against IEDs." Office of Public Affairs, February 27, 2007.

Burt, Chris. "Federal News Radio Explores U.S. Department of Defense 30-Year Roadmap for Biometrics." BiometricsUpdates.com, January 17, 2018.

"Can Math and Science Help Solve Crimes? Scientists Work with Los Angeles Police to Identify and Analyze Crime 'Hotspots.'" *Science Daily*, Science News, University of California, Los Angeles, February 27, 2010.

Canadian Press. "U.S. Troops Take Root in Former Canadian Base." CTV News, August 22, 2010.

Caterinicchia, Dan. "DOD Opens Biometrics Site." *Federal Computer Week*, July 10, 2003.

Cave, Damian. "250 Are Killed in Major Iraq Battle." *New York Times*, January 29, 2007.

Chandrasekaran, Rajiv. "In Afghanistan's South, Signs of Progress in Three Districts Signal a Shift." *Washington Post*, April 16, 2011.

Chen, Sophia. "This Drone Once Fought Wars. Now It's Fighting Climate Change." *Wired*, May 1, 2017.

"China: Minority Region Collects DNA from Millions, Private Information Gathered by Police, Under Guise of Public Health Program." Human Rights Watch, December 13, 2017.

CJIATF-435 Public Affairs, Combined Joint Interagency Task Force 435. "Rule of Law Makes Impressive Gains at Justice Center." U.S. Central Command. Camp Phoenix, Afghanistan, December 10, 2012.

"Clint Lorance LIVE in Studio." *The Sean Hannity Show*, November 18, 2019.

"CNBC Exclusive, CNBC Transcript: Palantir Technologies Cofounder & CEO Alex Karp Joins CNBC's Josh Lipton for a Rare

Interview Airing Today." CNBC News Releases, February 28, 2018.

"Congressional Hispanic Caucus Members Demand Trump Administration Release HHS Contracts with Palantir, Tech Companies." Congressional Hispanic Caucus press release, June 25, 2020.

Daugman, John. "Major International Deployments of the Iris Recognition Algorithms: 1.5 Billion Persons." University of Cambridge.

Davis, Ladonna S. "Team in Afghanistan Named Best Project Development Team in U.S. Army Corps of Engineers." U.S. Army Corps of Engineers, August 1, 2011.

Department of Homeland Security. "Snapshot: Rapid DNA Identifies Conception Boat Fire Victims." Newsroom, DHS.gov, December 10, 2019.

———. "Snapshot: S&T's Rapid DNA Technology Identified Victims of California Wildfire." Newsroom, DHS.gov, April 23, 2019.

DeJesus, Ivey. "Military Veteran Cpl. Matthew Hanes Honored and Remembered." Penn Live, August 14, 2015.

Demarest, Lt. Col. Geoffrey B., U.S. Army (ret.), PhD, JD, and Lt. Col. Lester W. Grau, U.S. Army (ret.). "Maginot Line or Fort Apache? Using Forts to Shape the Counterinsurgency Battle." *Military Review* (November–December 2005).

Dickerson, Caitlin. "U.S. Government Plans to Collect DNA from Detained Immigrants." *New York Times,* October 2, 2019.

Dolven, Taylor. "The U.S. Is Still Holding 'About 100' Kids Under 5 Who Were Taken from Parents at Border." Vice News, July 5, 2018.

Dongus, Ariana. "Galton's Utopia—Data Accumulation in Biometric Capitalism." *Spheres: Journal for Digital Cultures*, November 20, 2019.

"Dr. George Mohler: Mathematician and Crime Fighter." Data-Smart, City Solutions, Harvard Kennedy School Ash Center for Democratic Governance and Innovation, May 8, 2013.

Drew, Christopher. "Military Is Awash in Data from Drones." *New York Times*, January 10, 2010.

Drohan, Ed. "CSI Afghanistan: Forensic Experts Help Turn Bomb Maker into Convict." Defense Visual Information Distribution Service, December 10, 2013.

Faulds, Henry. "On the Skin-Furrows of the Hand." *Nature*, October 8, 1880.

"The FBI and the American Gangster, 1924–1938." Press conference, April 18, 1925, FBI.gov.

The FBI Laboratory 2006 Report, An FBI Laboratory Publication, Federal Bureau of Investigation, Quantico, VA.

FBI Law Enforcement Bulletin 69, no. 6 (June 2000).

"FBI Names New Laboratory Director: DiZinno Will Lead the Federal Bureau of Investigation Laboratory." FBI National Press Office, September 12, 2006.

Feliciano, Cpt. Juan P. "Training Your Company Intelligence-Support Team." *Armor* (April–June 2013): 31–33.

Ferguson, Andrew G. "Policing Predictive Policing." *Washington University Law Review* 94, no. 5 (2017).

Fidler, Steven. "Fears over Covert DNA Database." *Financial Times*, November 17, 2008.

"Fiftieth Anniversary of First Digital Image Marked." National Institute of Standards and Technology, May 24, 2007.

"Flushing Out Taliban." *The Heartbeat* 8, no. 4 (March 2011).

"Follow-up on the News; The 2 Identities of David Roberts." *New York Times*, September 25, 1988.

Fox, John F., Jr. "The Birth of the FBI's Technical Laboratory—1924 to 1935," FBI.gov.

Freudenrich, Craig, PhD. "How IEDs Work." HowStuffWorks: Science, December 10, 2008.

Garamone, Jim. "Petraeus Puts Protecting People at Strategy's Center." American Forces Press Service, August 3, 2010.

Gates, Robert. "Future Military Strategy." National Defense University, September 29, 2008.

Gavett, Gretchen. "What Is the Secretive US 'Kill/Capture' Campaign?" PBS *Frontline*, June 17, 2011.

Gilbert, Ben. "Afghanistan War: Bulldozing Through Kandahar." Public Radio International, November 10, 2010.

Gillert, Douglas J. "Who Are You? DNA Registry Knows." American Forces Press Service, July 13, 1998.

Goh, Melisa. "Former Captain: Afghan Shooting Suspect Showed 'Valorous Conduct' in Battle." National Public Radio, March 19, 2012.

Goldman, Adam. "Saudi Suspected of Waiting to Aid 9/11 Hijackers Seeks to Leave Guantanamo." *Washington Post*, June 14, 2016.

Goldstein, A. J., L. D. Harmon, and A. B. Lesk. "Identification of Human Faces." *Proceedings of the IEEE* 59, no. 5 (May 1971): 748–60.

Graves, Garret, Paul A. Gosar, Scott Perry, Neal P. Dunn, Duncan Hunter, Adam Kinzinger, Brian Babin, Michael Waltz. "Dear Mr. President . . ." Congress of the United States, Washington, DC, 20515, March 1, 2019. Author collection.

Greenberg, Andy. "How a 'Deviant' Philosopher Built Palantir, a CIA-Funded Data-Mining Juggernaut." *Forbes*, August 14, 2013.

Harris, Mark. "How Peter Thiel's Secretive Data Company Pushed into Policing." *Wired,* August 9, 2017.

Haskins, Caroline. "Academics Confirm Major Predictive Policing Algorithm Is Fundamentally Flawed." *Vice,* February 14, 2019.

Hersh, Seymour M. "Torture at Abu Ghraib." *The New Yorker,* May 10, 2004.

Himat, Abobakar, and Selim Dogan. "Ancient Karez System in Afghanistan: The Perspective of Construction and Maintenance." *Academic Platform Journal of Engineering and Science,* March 29, 2019.

Hu, Margaret. "Big Data Blacklisting." *Florida Law Review* 67, no. 5, Article 5 (March 2016): 1–76.

———. "Biometric Cyberintelligence and the Posse Comitatus Act." *Emory Law Journal* 66 (2017): 697–763.

Huet, Ellen. "Server and Protect: Predictive Policing Firm PredPol Promises to Map Crime Before It Happens." *Forbes,* February 11, 2015.

"Identity Dominance: The U.S. Military's Biometric War in Afghanistan." PublicIntelligence.net, April 21, 2014.

"In Memoriam, Woodrow Wilson Bledsoe." UTexas.edu.

International Campaign to Ban Landmines. "Landmine & Cluster Munition Monitor Fact Sheet." November 2011.

Kealy, Sean J. "Reexamining the Posse Comitatus Act: Toward a Right to Civil Law Enforcement." *Yale Law & Policy Review* 21, no. 2, Article 3 (2003).

Kemp, Kelly. "Clapper Inducted into the NGA Hall of Fame." *Pathfinder,* May 2008.

Khan, Waqar A. "Indelible Imprints: The Genius from Khulna." *Daily Sun,* May 8, 2017.

Kimery, Anthony. "DNA Processing Carried Out by DHS S&T for FBI, Intel Agencies." BiometricUpdate.com, January 14, 2019.

Krooks, David A., Lucy A. Whalley, and H. Garth Anderson. "Contingency Bases and the Problem of Sociocultural Context." U.S. Army Corps of Engineers, ERDC/CERL TN-12-2, July 2012.

Kurtzleben, Danielle. "CHART: How the U.S. Troop Levels in Afghanistan Have Changed Under Obama." National Public Radio, July 6, 2016.

Lamothe, Dan. "How 775,000 U.S. Troops Fought in One War: Afghanistan Military Deployments by the Numbers." *Washington Post,* September 11, 2019.

Lin, Liza, and Newley Purnell. "A World with a Billion Cameras Watching You Is Just Around the Corner." *Wall Street Journal,* December 6, 2019.

Magnuson, Stew. "Friend or Foe? Defense Department Under Pressure to Share Biometric Data." *National Defense Magazine,* January 1, 2009.

———. "Military 'Swimming in Sensors and Drowning in Data.'" *National Defense Magazine,* January 1, 2010.

Maher, John N. "Dear Mr. President, Mr. Pardon Attorney, and Mr. Secretary: On Behalf of First Lieutenant Clint A. Lorance." June 1, 2018. Author collection.

Maher, John N., and David Bolgiano. "Habeas Corpus Review of Military Convictions in the Federal Judiciary: An Enduring Need to Protect Constitutional Liberties." *Journal of Law, Policy & Military Affairs* 26 (2019): 1–51.

Maizland, Lindsay. "China's Repression of Uighurs in Xinjiang." Council on Foreign Relations, November 25, 2019.

Martins, Mark. "NATO Stands Up Rule of Law Field Support Mission in Afghanistan." *Lawfare Blog*, July 6, 2011.

McAleenan, Kevin K. et al. "Memorandum of Agreement Among the Office of Refugee Resettlement of the U.S. Department of Health and Human Services, and U.S. Immigration and Customs Enforcement and U.S. Customs and Border Protection of the U.S. Department of Homeland Security Regarding Consultation and Information Sharing Matters in Unaccompanied Alien Children Matters," April 13, 2018.

McClelland, Maj. Carol. "Improved Training Helps Forensics Team Prepare for Afghanistan Deployment." U.S. Army, March 24, 2011.

"Member of the Lorance Legal Team: We Need President Trump to Take Action, Donald Brown Says Army 1st Lt. Col. Clint Lorance Is Not Guilty of War Crimes." *Fox & Friends*, April 13, 2019.

Miroff, Nick. "Homeland Security to Test DNA of Families at Border in Cases of Suspected Fraud." *Washington Post*, May 1, 2019.

Morris, Victor R. "Why COIST Matters." *Small Wars Journal*, 2015.

Moruza, Paul. "Intelligence Center Develops Biometrically Enabled Intelligence to Support Warfighter." Army.mil, January 9, 2013.

Moses, Kenneth R. et al. "Automated Fingerprint Identification System (AFIS)." National Institute of Justice, *Fingerprint Sourcebook*, 6:3–5.

Munn, Nathan. "This Predictive Policing Company Compares Its Software to 'Broken Windows' Policing." Motherboard: Tech by *Vice*, June 11, 2018.

Nadem, Bashir Ahmad. "21 Killed, 50 Injured in Twin Suicide Blasts." *Pajhwok Afghan News*, June 6, 2012.

"Name Index to Leavenworth Federal Penitentiary Inmate Case Files, 1895–1931." National Archives at Kansas City, MO, Archives.gov.

"Nato Apologises for Afghan Koran 'Burning.'" BBC Asia, February 21, 2012.

Nordland, Rod. "3 NATO Soldiers Killed by Afghan Security Officers." *New York Times,* March 26, 2012.

———. "Afghanistan Has Big Plans for Biometric Data." *New York Times,* November 19, 2011.

Obama, Barack. "National Security Strategy." The White House, May 2010.

Olsen, Robert D., Sr. "A Fingerprint Fable: The Will and William West Case." *Identification News* 37, no. 11 (November 1987).

Pendall, David, and Cal Sieg. "Biometric-enabled Intelligence in Regional Command–East." *Joint Force Quarterly* 72 (1st Quarter 2014): 69–74.

Petit, Lt. Col. Brian. "The Fight for the Village: Southern Afghanistan 2010." *Military Review* (May–June 2011): 25–32.

Petraeus, Gen. David H. "COMISAF's Counterinsurgency Guidance." Headquarters, International Security Assistance Force/United States Forces–Afghanistan, August 1, 2010.

Philipps, Dave. "Cause Célèbre, Scorned by Troops." *New York Times,* February 24, 2015.

Phillips, P. Jonathon et al. "Face Recognition Accuracy of Forensic Examiners, Superrecognizers, and Face Recognition Algorithms." *PNAS* 114, no. 24 (June 12, 2018).

Phillips, Stephen. "Birth of the Combined Explosives Exploitation Cell." *Small Wars Journal* (2008).

Press, Margaret, PhD, and Colleen Fitzpatrick, PhD. "The DNA DOE Project, Our First Two Years of Success." Presentation, International Symposium on Human Identification, September 26, 2019.

Rahimi, Sangar, and Alissa J. Rubin. "Koran Burning in NATO Error Incites Afghans." *New York Times,* February 21, 2012.

Raiten, Jesse M., MD. "Among Body Parts and Colleagues: Finding My Team in the Rubble on 9/11." *Anesthesiology* 129 (December 2018): 1186–188.

Ramsey, John. "Army First Lieutenant Found Guilty of Murder, Other Charges for Actions in Afghanistan." *Fayetteville Observer,* August 2, 2013.

Ransom, Jan, and Ashley Southall. "N.Y.P.D. Detectives Gave a Boy, 12, a Soda. He Landed in a DNA Database." *New York Times,* August 15, 2019.

"Raytheon to Supply Biometrics to DOD." Raytheon press release, January 28, 2009.

Reel, Monte. "Secret Cameras Watch Baltimore's Every Move from Above." *Bloomberg Businessweek,* August 23, 2016.

Reinert, Patrick J., and John F. Hussey. "The Military's Role in Rule of Law Development." *Joint Force Quarterly* 77 (2nd Quarter 2015): 120–127.

"Rise of the Drones." PBS *Nova,* January 23, 2013.

Roggio, Bill. "Taliban Kill 21 Afghans in Double Suicide Attack." *Long War Journal,* June 6, 2012.

Rubin, Alissa J. "Chain of Avoidable Errors Cited in Koran Burning." *New York Times,* March 2, 2012.

Rubin, Alissa J., and Taimoor Shah. "Afghanistan Faces Deadliest Day for Civilians This Year in Multiple Attacks." *New York Times,* June 6, 2012.

"Rule of Law in Afghanistan Is Critical to an Enduring Transition of Governance, Says HLS Medal of Freedom Recipient Brig. Gen. Mark Martins '90." *Harvard Law Today,* May 22, 2011.

"Rule of Law Programs in Afghanistan." Department of State, Fact Sheet, May 4, 2012.

Rumsfeld, Donald, to Gen. Dick Myers, Paul Wolfowitz, Gen. Pete Pate, and Doug Feith. "Global War on Terrorism." Memo, October 16, 2003.

Saini, Monika, and Anup Kumar Kapoor. "Biometrics in Forensic Identification: Applications and Challenges." *Journal of Forensic Medicine* 1, no. 2 (May 2016).

Sandoval, Edgar. "N.Y.P.D. to Remove DNA Profiles of Non-Criminals from Database." *New York Times*, February 20, 2020.

Shannon, Paul J. "Fingerprints and the War on Terror: An FBI Perspective." *Joint Force Quarterly* 43 (4th Quarter 2006): 76–82.

Short, Martin B., P. Jeffrey Brantingham, Andrea L. Bertozzi, and George E. Tita. "Dissipation and Displacement of Hotspots in Reaction-Diffusion Models of Crime." *PNAS*, February 22, 2010.

Smith, Thomas B., and Marc Tranchemontagne. "Understanding the Enemy: The Enduring Value of Technical and Forensic Exploitation." *Joint Force Quarterly* 75 (4th Quarter 2014): 122–28.

Southerland, Vincent. "With AI and Criminal Justice, the Devil Is in the Data." ACLU.org, April 9, 2018.

Special Operations Forces Acquisition, Technology, and Logistics, Directorate of Science and Technology (SOF AT&L-ST), Broad Agency Announcement, USSOCOM-BAAST-2020, unclassified, 4.0-4.1.1.1.2.

Statement of Jose E. Melendez-Perez to the National Commission on Terrorist Attacks upon the United States, Hart Senate Office Building, Washington, DC, January 26, 2004.

"Status of IDENT/IAFIS Integration," Report No. I-2002-003, December 7, 2001, USDOJ.org.

Sussman, Phil. "COIST Staffs Play Crucial Role on Today's Complex Battlefield." U.S. Army press release, June 19, 2009.

Timberg, Craig. "New Surveillance Technology Can Track Everyone in an Area for Several Hours at a Time." *Washington Post,* February 5, 2014.

Tita, Dr. George E. "Human Terrain Mapping, Geospatial Intelligence, LAPD—DGI." University of California, Irvine, transcript, February 3, 2012.

"Unclassified Summary of Final Determination." Periodic Review Board, Detainee Name: Mohammed Mani Ahmad al-Qahtani, Detainee ISN 63, July 18, 2016.

U.S. Central Command. "Conference Maps the Way Ahead for Biometrics in Afghanistan." Kabul, Afghanistan, October 15, 2010.

U.S. Department of Defense, Office of the Secretary, Defense Science Board. "Action: Notice of Advisory Committee Meeting." *Federal Register* 69, no. 26, Monday, February 9, 2004.

Vaughan, Brendan. "Robert Bales Speaks: Confessions of America's Most Notorious War Criminal." *GQ*, October 21, 2015.

"Visible Proofs: Forensic Views of the Body." National Library of Medicine. NLM.NIH.gov, 2008.

Wee, Sui-Lee. "China Uses DNA to Track Its People, with the Help of American Expertise." *New York Times,* February 21, 2019.

Weir, Dr. Gary E. "The Evolution of Geospatial Intelligence and the National Geospatial-Intelligence Agency." *Intelligencer: Journal of U.S. Intelligence Studies* 21, no. 3 (Fall/Winter 2015): 53–59.

"West Virginia on Cutting Edge of Latest Advances in Biometrics." *Capacity*, Fall 2006.

Woodward, John D., Jr. "How Do You Know Friend from Foe?" *Homeland Science and Technology*, December 2004.

———. "Using Biometrics to Achieve Identity Dominance in the Global War on Terrorism." U.S. Army Combined Arms Center, September 2005.

"Working Through the Claims Process in Zharay." *The Heartbeat* 8, no. 4 (March 2011).

Wu, Nicholas, and John Fritze. "Trump Pardons Service Members in High Profile War Crimes Cases." *USA Today,* November 15, 2019.

Yager, MCC (SW) Maria. "Governors' Visit to DFIP Provides a Firsthand Look at Detention Operations." U.S. Central Command. Parwan Province, Afghanistan, February 2, 2011.

"Your Right to Federal Records: Questions and Answers on the Freedom of Information Act and the Privacy Act, 1992." Electronic Privacy Information Center, EPIC.org.

COURT DOCUMENTS, LEGAL REPORTS, AND TRANSCRIPTS

Affidavit of William C. Carney. State of Illinois, County of Cook, August 31, 2015.

Allen, Gen. John. R. "Report of Investigation IAW AR 15-6 Facts and Circumstances Surrounding Allegations of Shooting Afghan Civilians Outside Village Stability Platform Belambai," August 13, 2015.

Committee on Armed Services. "Department of Defense Authorization for Appropriations for Fiscal Year 2007." Volume 4, Part 1, February–March 2006.

McGehee v. C.I.A., 697 F.2d, 1095, U.S. District Court of Appeals, District of Columbia Circuit, October 4, 1983.

Supreme Court of the United States. *Maryland v. King.* Certiorari to the Court of Appeals of Maryland. No 12-207. October Term, 2012.

———. *Robert Bales v. United States.* On Petition for a Writ of Certiorari to the United States Court of Appeals for the Armed Forces. No. 17. Washington, DC, May 16, 2018.

The United States v. First Lieutenant Clint Lorance. Proceedings of a General Court-Martial. Fort Bragg, NC, April 25, 2013; July 30, 2013.

U.S. Army, Criminal Investigation Command. *Commander's Report of Disciplinary or Administrative Action: LORANCE, Clint Allen.* AR 190-45/AR 195-2. Crime Records Center, Quantico, VA, August 7, 2013.

———. "All Images (CRC): Exhibit 124." Crime Records Center, Quantico, VA, September 7, 2013: 000482-000572.

———. *Summary of Investigative Activity: Lt Lorance.* Control No. 0254-2012-CID379-77688. 2013.

U.S. Court of Appeals for the Armed Forces. *The United States v. Lt. Clint A. Lorance, Supplement to Petition for Grant or Review.* USCA Dkt. No. 17-0599/AR. October 10, 2011.

U.S. District Court, District of Columbia. *Clint A. Lorance v. Department of the Army.* Civil Action No. 18-1710 (BAH). August 8, 2019.

U.S. District Court, District of Kansas. *Clint A. Lorance v. Commandant.* Declaration of Kevin H. Case No. 18-3297-JWL. May 19, 2019.

SLIDE DECKS

Biometrics Task Force 19. *Biometric Automated Toolset (BAT) and Handheld Interagency Identity Detection Equipment (HIIDE).* Overview for NIST XML & Mobile ID Workshop. September 2007.

Chelko, Larry, C. *Department of Defense Forensic Capabilities*. U.S. Army Criminal Investigation Laboratory (USACIL) (n.d.).

Dee, Tom. *NDIA: Disruptive Technologies: U.S. Army Intelligence Center of Excellence*. U.S. Department of Defense, Biometrics. September 5, 2007.

Devabhakthuni, Bharatha. *Biometric Interoperability*. Criminal Justice Information Services Division, Interoperability Initiatives Unit. November 2, 2011.

National Conference of State Legislatures. *DNA Arrestee Laws*. 2013.

National Geospatial-Intelligence Agency. *Activity Based Intelligence*. Approved for Public Release 13-231 (n.d.).

New Systems Training and Integration Office. *Introduction to Biometrics and Biometric Systems*. Biometrics Identity and Management Agency. July 7, 2011. Author collection.

"TEDAC Marks 10-Year Anniversary: A Potent Weapon in the War on Terror." News, FBI.gov, December 12, 2013.

Tontarski, Rick. *Defense Forensic Enterprise System*. DoD Forensics Workshop. U.S. Army Criminal Investigation Command and Criminal Investigation Laboratory. The National Academies, Needs of the Forensic Science Community, Woods Hole, MA, September 21, 2007.

U.S. Army Criminal Investigation Command. *U.S. Army Criminal Investigation Command*. Quantico, VA, 2019.

INDEX

INDEX

ABOUT THE AUTHOR

Annie Jacobsen is the author of the Pulitzer Prize finalist in history *The Pentagon's Brain*, the *New York Times* bestsellers *Area 51* and *Operation Paperclip*, and other books. She was a contributing editor at the *Los Angeles Times Magazine.* A graduate of Princeton University, she lives in Los Angeles with her husband and two sons.